CAMBRIDGE LIBRARY COLLECTION

Books of enduring scholarly value

Spiritualism and Esoteric Knowledge

Magic, superstition, the occult sciences and esoteric knowledge appear regularly in the history of ideas alongside more established academic disciplines such as philosophy, natural history and theology. Particularly fascinating are periods of rapid scientific advances such as the Renaissance or the nineteenth century which also see a burgeoning of interest in the paranormal among the educated elite. This series provides primary texts and secondary sources for social historians and cultural anthropologists working in these areas, and all who wish for a wider understanding of the diverse intellectual and spiritual movements that formed a backdrop to the academic and political achievements of their day. It ranges from works on Babylonian and Jewish magic in the ancient world, through studies of sixteenth-century topics such as Cornelius Agrippa and the rapid spread of Rosicrucianism, to nineteenth-century publications by Sir Walter Scott and Sir Arthur Conan Doyle. Subjects include astrology, mesmerism, spiritualism, theosophy, clairvoyance, and ghost-seeing, as described both by their adherents and by sceptics.

The Night Side of Nature

The novelist and children's author Catherine Crowe (c.1800–76) published *The Night Side of Nature* in two volumes in 1848. This lively collection of ghostly sketches and anecdotes was a Victorian best-seller and Crowe's most popular work. Sixteen editions appeared in six years, and it was translated into several European languages. The stories are intertwined with Crowe's own interpretations and commentaries which attack the scepticism of enlightenment thought and orthodox religion. Crowe seeks instead to encourage and re-invigorate a sense of wonder and mystery in life by emphasising the supernatural. Volume 2 probes the mysterious phenomena of troubled spirits, haunted houses, spectral lights, apparitions and poltergeists. Crowe's vivid tales, written with great energy and imagination, are classic examples of nineteenth-century spiritualist writing and strongly influenced other authors, including Charles Baudelaire, as well as providing inspiration for later adherents of ghost-seeing and psychic culture.

T0273856

Cambridge University Press has long been a pioneer in the reissuing of out-of-print titles from its own backlist, producing digital reprints of books that are still sought after by scholars and students but could not be reprinted economically using traditional technology. The Cambridge Library Collection extends this activity to a wider range of books which are still of importance to researchers and professionals, either for the source material they contain, or as landmarks in the history of their academic discipline.

Drawing from the world-renowned collections in the Cambridge University Library, and guided by the advice of experts in each subject area, Cambridge University Press is using state-of-the-art scanning machines in its own Printing House to capture the content of each book selected for inclusion. The files are processed to give a consistently clear, crisp image, and the books finished to the high quality standard for which the Press is recognised around the world. The latest print-on-demand technology ensures that the books will remain available indefinitely, and that orders for single or multiple copies can quickly be supplied.

The Cambridge Library Collection will bring back to life books of enduring scholarly value (including out-of-copyright works originally issued by other publishers) across a wide range of disciplines in the humanities and social sciences and in science and technology.

The Night Side of Nature

Or, Ghosts and Ghost Seers

VOLUME 2

CATHERINE CROWE

CAMBRIDGE
UNIVERSITY PRESS

CAMBRIDGE UNIVERSITY PRESS

Cambridge, New York, Melbourne, Madrid, Cape Town, Singapore,
São Paolo, Delhi, Dubai, Tokyo, Mexico City

Published in the United States of America by Cambridge University Press, New York

www.cambridge.org
Information on this title: www.cambridge.org/9781108027502

© in this compilation Cambridge University Press 2011

This edition first published 1848
This digitally printed version 2011

ISBN 978-1-108-02750-2 Paperback

THE

NIGHT SIDE OF NATURE;

OR,

GHOSTS AND GHOST SEERS.

BY

CATHERINE CROWE,

AUTHORESS OF

"SUSAN HOPLEY," "LILLY DAWSON,"
" ARISTODEMUS," &c. &c.

" Thou com'st in such a questionable shape,
That I will speak to thee!"
HAMLET.

IN TWO VOLUMES.

VOL. II.

LONDON:
T. C. NEWBY, 72, MORTIMER St., CAVENDISH Sq.

1848.

INDEX.

VOL. II.

THE

NIGHT SIDE OF NATURE.

CHAPTER I.

THE POWER OF WILL

THE power, be it what it may, whether of dressing up an ethereal visible form, or of acting on the constructive imagination of the seer, which would enable a spirit to appear " in his habit as he lived," would also enable him to present any other object to the eye of the seer, or himself in any shape, or fulfilling any function he willed; and we thus find in various instances, especially those recorded in the Seeress of Prevorst, that this is the case. We not only see changes of dress, but we see books, pens, writing materials, &c. in their hands; and we find a great variety of sounds imitated : which sounds are frequently heard, not only by those who have the faculty of " discerning of spirits," as St. Paul says, but also by every other person on the spot; for the hearing these sounds does not seem to depend on any particular faculty on the part of the auditor, except it be in the case of speech. The hearing the speech of a spirit, on the contrary, appears in most instances to be dependant on the same conditions as the seeing it, which may possibly arise from there being, in fact, no *audible* voice at all, but the same sort of spiritual communication which exists between a magnetiser and his patient, wherein the sense is conveyed without words.

This imitating of sounds, I shall give several instances of in a future chapter. It is one way in which a death is frequently indicated. I could quote a number of examples of this description, but shall confine myself to two or three.

Mrs. D. being, one night, in her kitchen, preparing to go to bed, after the house was shut up and the rest of the family retired, she was startled by hearing a foot coming along the passage, which she recognised distinctly to be that of her father, who she was quite certain was not in the house. It advanced to the kitchen-door, and she waited with alarm to see if the door was to open ; but it did not, and she heard nothing more. On the following day, she found that her father had died at that time ; and it was from her niece I heard the circumstance.

A Mr. J. S., belonging to a highly respectable family, with whom I am acquainted, having been for some time in declining health, was sent abroad for change of air. During his absence, one of his sisters, having been lately confined, an old servant of the family was sitting half asleep in an arm-chair, in a room adjoining that in which the lady slept, when she was startled by hearing the foot of Mr. J. S. ascending the stairs. It was easily recognisable, for, owing to his constant confinement to the house, in consequence of his infirm health, his shoes were always so dry that their creaking was heard from one end of the house to the other. So far surprised out of her recollection as to forget he was not in the country, the good woman started up, and, rushing out with her candle in her hand, to light him, she followed the steps up to

Mr. J. S.'s own bed-chamber, never discovering that he was not preceding her till she reached the door. She then returned, quite amazed, and, having mentioned the occurrence to her mistress, they noted the date, and it was afterwards ascertained that the young man had died at Lisbon on that night.

Mrs. F. tells me that, being one morning, at eleven o'clock, engaged in her bed-room, she suddenly heard a strange, indescribable, sweet, but unearthly sound, which apparently proceeded from a large open box which stood near her. She was seized with an awe and a horror which there seemed nothing to justify, and fled up stairs to mention the circumstance, which she could not banish from her mind. At that precise day and hour, eleven o'clock, her brother was drowned. The news reached her two days afterwards.

Instances of this kind are so well known, that it is unnecessary to multiply them further. With respect to the mode of producing these sounds, however, I should be glad to say something more definite if I could, but, from the circumstance of their being heard not only by one person, who might be supposed to be *en rapport*, or whose constructive imagination might be acted upon, but by any one who

happens to be within hearing, we are led to conclude that the sounds are really reverberating
through the atmosphere. In the strange cases
recorded in "The Seeress of Prevorst," although
the apparitions were visible only to certain
persons, the sounds they made were audible to
all; and the seeress says they are produced by
means of the nerve-spirit, which I conclude is
the spiritual body of St. Paul and the atmosphere, as we produce sound by means of
our *material* body and the atmosphere.

In this plastic power of the spirit to present
to the eye of the seer whatever object it wills,
we find the explanation of such stories as the
famous one of Ficinus and Mercatus, related
by Baronius in his annals. These two
illustrious friends, Michael Mercatus and
Marcellinus Ficinus, after a long discourse on
the nature of the soul, had agreed that, if possible, whichever died first should return to
visit the other. Some time afterwards, whilst
Mercatus was engaged in study at an early
hour in the morning, he suddenly heard the
noise of a horse galloping in the street, which
presently stopped at his door, and the voice of
his friend Ficinus exclaimed, "Oh, Michael!
Oh, Michael! *Vera sunt illa!* Those things

are true!" Whereupon Mercatus hastily opened his window, and espied his friend Ficinus on a white steed. He called after him, but he galloped away out of his sight. On sending to Florence to enquire for Ficinus, he learnt that he had died about that hour he called to him. From this period to that of his death, Mercatus abandoned all profane studies, and addicted himself wholly to Divinity. Baronius lived in the sixteenth century, and even Dr. Ferrier and the spectral illusionists admit that the authenticity of this story cannot be disputed, although they still claim it for their own.

Not very many years ago, Mr. C., a staid citizen of Edinburgh, whose son told me the story, was one day riding gently up Corstorphina-hill, in the neighbourhood of the city, when he observed an intimate friend of his own, on horseback also, immediately behind him, so he slackened his pace to give him an opportunity of joining company. Finding he did not come up so quickly as he should, he looked round again, and was astonished at no longer seeing him, since there was no side road into which he could have disappeared. He returned home, perplexed at the oddness of the circumstance, when the first thing he learnt was, that, during his absence, this friend had been killed by his horse falling in Candlemaker's-row.

I have heard of another circumstance, which occurred some years ago in Yorkshire, where, I think, a farmer's wife was seen to ride into a farm-yard on horse-back, but could not be afterwards found, or the thing accounted for, till it was ascertained that she had died at that period.

There are very extraordinary stories extant in all countries of persons being annoyed by appearances in the shape of different animals, which one would certainly be much disposed to give over altogether to the illusionists, though at the same time it is very difficult to reduce some of the circumstances under that theory; especially, one mentioned, p. 307, of my "Translation of the Seeress of Prevorst." If they are not illusions, they are phenomena to be attributed, either to this plastic power, or to that magico-magnetic influence in which the belief in lycanthropy and other strange transformations have originated. The multitudes of unaccountable stories of this description recorded in the witch trials, have long furnished a subject of perplexity to everybody who was sufficiently just to human nature to

conclude, that there must have been some
strange mystery at the bottom of an infatu-
ation that prevailed so universally, and in
which so many sensible, honest, and well-
meaning persons were involved. Till of late
years, when some of the arcana of animal or
vital magnetism have been disclosed to us, it
was impossible for us to conceive by what
means such strange conceptions could prevail;
but since we now know, and many of us have
witnessed, that all the senses of a patient are
frequently in such subjection to his magnetiser,
that they may be made to convey any impressions to the brain that magnetiser wills, we
can, without much difficulty, conceive how this
belief in the power of transformation took its
rise; and we also know how a magician could
render himself visible or invisible, at pleasure.
I have seen the sight or hearing of a patient
taken away and restored, by Mr. Spencer Hall,
in a manner that could leave no doubt on the
mind of the beholder; the evident paralysis
of the eye of the patient testifying to the fact.
Monsieur Eusèbe Salverte, the most determined of rationalistic sceptics, admits that we
have numerous testimonies to the existence
of an art, which he confesses himself at some
loss to explain, although the opposite quarters

from which the accounts of it reach us, render
it difficult to imagine that the historians have
copied each other. The various transformations of the gods into eagles, bulls, and so
forth, have been set down as mere mytho-
logical fables ; but they appear to have been
founded on an art, known in all quarters of
the world, which enabled the magician to
take on a form that was not his own, so as to
deceive his nearest and dearest friends. In
the history of Gengis Khan there is mention
of a city, which he conquered, "in which
dwelt," says Suidas, "certain men, who possessed the secret of surrounding themselves
with deceptive appearances, insomuch that
they were able to represent themselves to the
eyes of people quite different to what they
really were." Saxo Grammaticus, in speaking
of the traditions connected with the religion
of Odin, says, "that the magi were very
expert in the art of deceiving the eyes, being
able to assume, and even to enable others
to assume, the forms of various objects, and
to conceal their real aspects under the most
attractive appearances." John of Salisbury,
who seems to have drawn his information from
sources now lost, says, that " Mercury, the
most expert of magicians, had the art of fasci-

nating the eyes of men to such a degree, as to render people invisible, or make them appear in forms quite different to those they really bore. We also learn, from an eye-witness, that Simon, the magician, possessed the secret of making another person resemble him so perfectly, that every eye was deceived. Pomponius Mela affirms, that the Druidesses of the island of Sena could transform themselves into any animal they chose, and Proteus has become a proverb by his numerous metamorphoses.

Then, to turn to another age and another hemisphere, we find Joseph Acosta, who resided a long time in Peru, assuring us that there existed at that period magicians who had the power of assuming any form they chose. He relates that the predecessor of Montezuma, having sent to arrest a certain chief, the latter successively transformed himself into an eagle, a tiger, and an immense serpent; and so eluded the envoys, till having consented to obey the king's mandate, he was carried to court and instantly executed.

The same perplexing exploits are confidently attributed to the magicians of the West Indies; and there were two men eminent amongst the natives, the one called Gomez and the other Gonzalez, who possessed this art in an extraordinary degree; but both fell victims to the practice of it, being shot during the period of their apparent transformations.

It is also recorded that Nanuk, the founder of the Sikhs, who are not properly a nation but a religious sect, was violently opposed by the Hindoo zealots; and at one period of his career when he visited Vatala, the Yogiswaras, who were recluses, that by means of corporeal mortifications, were supposed to have acquired a command over the powers of nature, were so enraged against him, that they strove to terrify him by their enchantments, assuming the shapes of tigers and serpents. But they could not succeed, for Nanuk appears to have been a real philosopher, who taught a pure theism, and inculcated universal peace and toleration. His tenets, like the tenets of the founders of all religions, have been since corrupted by his followers. We can scarcely avoid concluding that the power by which these feats were performed is of the same nature as that by which a magnetiser persuades his patient that the water he drinks is beer, or the beer wine, and the analogy betwixt it, and that by which I have supposed a spirit to present himself, with such accompaniments as

he desires, to the eye a spectator, is evident.
In those instances where female figures are
seen with children in their arms, the appearance of the child we must suppose to be produced in this manner.

Spirits of darkness, however, cannot, as I
have before observed, appear as spirits of
light; the moral nature cannot be disguised.
On one occasion, when Frederica Hauffe asked
a spirit, if he could appear in what form he
pleased; he answered, no; that if he had lived
as a brute, he should appear as a brute; " as
our dispositions are, so we appear to you."

This plastic power is exhibited in those
instances I have related, where the figure has
appeared dripping with water, indicating the
kind of death that had been suffered; and
also in such cases as that of Sir Robert H. E.,
where the apparition showed a wound in his
breast. There are a vast number of similar
ones on record, in all countries; but I will
here mention one which I received from the
lips of a member of the family concerned,
wherein one of the trivial actions of life was
curiously represented.

Miss L. lived in the country with her three
brothers, to whom she was much attached, as
they were to her. These young men, who

amused themselves all the morning with their
out-door pursuits, were in the habit of coming
to her apartment, most days before dinner
and conversing with her till they were summoned to the dining-room. One day, when
two of them had joined her, as usual, and they
were chatting cheerfully over the fire, the door
opened, and the third came in, crossed the
room, entered an adjoining one, took off his
boots, and then instead of sitting down beside
them, as usual, passed again through the room,
went out, leaving the door open, and they saw
him ascend the stairs towards his own chamber, whither they concluded he was gone to
change his dress. These proceedings had been
observed by the whole party: they saw him
enter, saw him take off his boots, saw him
ascend the stairs, continuing the conversation
without the slightest suspicion of anything
extraordinary. Presently afterwards the dinner
was announced; and as this young man did
not make his appearance, the servant was desired to let him know they were waiting for
him. The servant answered, that he had not
come in yet; but being told that he would find
him in his bed-room, he went up stairs to call
him. He was however not there, nor in the
house; nor were his boots to be found where

he had been seen to take them off. Whilst
they were yet wondering what could have become of him, a neighbour arrived to break the
news to the family, that their beloved brother
had been killed whilst hunting, and that the
only wish he expressed was, that he could live
to see his sister once more.

I observed in a former chapter whilst speaking of wraiths, how very desirable it would be
to ascertain whether the phenomenon takes
place before, or after, the dissolution of the
bond betwixt soul and body; I have since
received the most entire satisfaction on that
head, so far as the establishing the fact, that it
does sometimes occur after the dissolution.
Three cases have been presented to me from
the most undoubted authority, in which the
wraith was seen at intervals varying from one
to three days after the decease of the person whose image it was; very much com-
plicating the difficulty of that theory which
considers these phenomena the result of an
interaction, wherein the vital principle of one
person is able to influence another within
its sphere, and thus make the organs of that
other the subjects of its will; a magical power,
by the way, which far exceeds that we possess
over our own organs. There is here, however,

where death has taken place, no living
organism to produce the effect, and the phenomenon becomes, therefore, purely subjective
—a mere spectral illusion, attended by a coin-
cidence, or else the influence is that of the
disembodied spirit, and those who will take
the trouble of investigating this subject, will
find that the number of these coincidences
would violate any theory of probabilities, to
a degree that precludes the acceptance of that
explanation. I do not see, therefore, on what
we are to fall back, except it be the willing
agency of the released spirit, unless we suppose
that the operation of the will of the dying
person travelled so slowly, that it did not take
effect till a day or two after it was exerted, an
hypothesis too extravagant to be admitted.

Dr. Passavent, whose very philosophical
work on this occult department of nature, is
well worth attention, considers the fact of these
appearances far too well established to be
disputed; and he enters into some curious
disquisitions with regard to what the Germans
call *far-working*, or the power of acting on
bodies at a distance, without any sensible
conductor, instancing the case of a gymnotus,
which was kept alive for four months in Stock-
holm, and which, when urged by hunger, could

kill fish at a distance, without contact, adding,
that it rarely miscalculated the amount of the
shock necessary to its purpose. These, and all
such effects, are attributed by this school of
physiologists to the supposed imponderable,
the nervous ether I have elsewhere mentioned, which Dr. P. conceives in cases of
somnambulism, certain sicknesses, and the
approach of death, to be less closely united
to its material conductors, the nerves, and
therefore capable of being more or less
detached, and acting at a distance, especially
on those with whom relationship, friendship,
or love, establish a rapport, or polarity ; and
he observes that intervening substances, or
distance, can no more impede this agency than
they do the agency of mineral magnetism.
And he considers that we must here seek for
the explanation of those curious so-called
coincidences of pictures falling, and clocks and
watches stopping, at the moment of a death,
which we frequently find recorded.

With respect to the wraiths, he observes
that the more the ether is freed, as by trance, or
the immediate approach of death, the more
easily the soul sets itself in rapport with distant persons ; and that thus it either acts
magically, so that the seer perceives the real

actual body of the person that is acting upon
him, or else that he sees the ethereal body,
which presents the perfect form of the fleshly
one, and which, whilst the organic life pro-
ceeds, can be momentarily detached and appear elsewhere, and this ethereal body he holds
to be the fundamental form, of which the
external body is only the copy, or husk.

I confess I much prefer this theory of Dr.
Passavent's, which seems to me to go very
much to the root of the matter. We have
here the " spiritual body" of St. Paul, and the
"nerve spirit" of the somnambulists, and its
magical effects are scarcely more extraordinary,
if properly considered, than its agency on our
own *material* bodies. It is this ethereal body
which obeys the intelligent spirit within, and
which is the intermediate agent betwixt the
spirit and the fleshly body. We here find the
explanation of wraiths, whilst persons are in
trance, or deep sleep, or comatose, this ethereal
body can be detached and appear elsewhere ;
and I think there can be no great difficulty
for those who can follow us so far, to go a
little further, and admit that this ethereal body
must be indestructible, and survive the death
of the material one ; and that it may, therefore, not only become visible to us under

given circumstances, but that it may, also, produce effects bearing some similarity to those it was formerly capable of, since, in acting on our bodies during life, it is already acting on a material substance, in a manner so incomprehensible to us, that we might well apply the word *magical,* when speaking of it, were it not that custom has familiarised us to the marvel.

It is to be observed, that this idea of a spiritual body is one that pervaded all Christendom, in the earlier and purer ages of Christianity, before priestcraft—and by priestcraft, I mean the priestcraft of all denominations—had overshadowed and obscured, by their various sectarian heresies, the pure teaching of Jesus Christ.

Dr. Ennemoser mentions a curious instance of this *actio in distans,* or far-working. It appears that Van Helmont having asserted that it was possible for a man to extinguish the life of an animal by the eye alone (*oculis intentis*), Rousseau, the naturalist, repeated the experiment, when in the East, and in this manner killed several toads; but on a subsequent occasion, whilst trying the same experiment at Lyons, the animal, on finding it could not escape, fixed its eyes immoveably on him, so that he fell into a fainting fit, and was thought to be dead. He was restored by means of theriacum and viper powder—a truly homeopathic remedy! However, we here probably see the origin of the universal popular persuasion, that there is some mysterious property in the eye of a toad; and also of the, so called, superstition of the *evil eye*.

A very remarkable circumstance occurred some years ago, at Kirkcaldy, when a person, for whose truth and respectability I can vouch, was living in the family of a Colonel M., at that place. The house they inhabited, was at one extremity of the town, and stood in a sort of paddock. One evening, when Colonel M. had dined out, and there was nobody at home but Mrs. M., her son, (a boy about twelve years old) and Ann, the maid, (my informant), Mrs. M. called the latter, and directed her attention to a soldier, who was walking backwards and forwards in the drying ground, behind the house, where some linen was hanging on the lines. She said, she wondered what he could be doing there, and bade Ann fetch in the linen, lest he should purloin any of it. The girl fearing he might be some ill-disposed person, felt afraid; Mrs. M., however, promising to watch from the window, that nothing happened to her, she went; but still apprehensive of the man's intentions, she turned her back towards him, and hastily pulling down the linen, she carried it into the house; he, continuing his walk the while, as before, taking no notice of her, whatever. Ere long, the Colonel returned, and Mrs. M. lost no time in taking him to the window to look at the man, saying, she could not conceive what he could mean, by walking backwards and forwards there, all that time ; whereupon, Ann added, jestingly, " I think, it's a ghost, for my part !" Colonel M. said, " he would soon see that," and calling a large dog that was lying in the room, and accompanied by the little boy, who begged to be permitted to go also, he stept out and approached the stranger ; when, to his surprise, the dog, which was an animal of high courage, instantly flew back, and sprung through the glass door, which the Colonel had closed behind him, shivering the panes all around.

The Colonel, meantime, advanced and challenged the man, repeatedly, without obtaining any answer or notice whatever; till, at length, getting irritated, he raised a weapon with which he had armed himself, telling him he " must speak, or take the consequences," when just as he was preparing to strike, lo! there was nobody there! The soldier had disappeared, and the child sunk senseless to the ground. Colonel M. lifted the boy in his arms, and as he brought him into the house, he said to the girl, "You are right, Ann. It *was* a ghost!" He was exceedingly impressed with this circumstance, and much regretted his own behaviour, and also the having taken the child with him, which he thought had probably prevented some communication that was intended. In order to repair, if possible, these errors, he went out every night, and walked on that spot for some time, in hopes the apparition would return. At length, he said, that he had seen and had conversed with it; but the purport of the conversation he would never communicate to any human being; not even to his wife. The effect of this occurrence on his own character was perceptible to everybody that knew him. He became grave and thoughtful, and appeared like one who had passed through some strange experience. The above-named Ann H., from whom I have the account, is now a middle-aged woman. When the circumstance occurred, she was about twenty years of age. She belongs to a highly respectable family; and is, and always has been, a person of unimpeachable character and veracity.

In this instance, as in several others I meet with, the animal had a consciousness of the nature of the appearance, whilst the persons around him had no suspicion of anything unusual. In the following singular case, we must conclude that attachment counteracted this instinctive apprehension. A farmer, in Argyleshire, lost his wife, and a few weeks after her decease, as he and his son were crossing a moor, they saw her sitting on a stone, with their house-dog lying at her feet, exactly as he used to do when she was alive. As they approached the spot the woman vanished, and supposing the dog must be equally visionary, they expected to see him vanish, also ; when, to their surprise, he rose and joined them, and they found it was actually the very animal of flesh and blood. As the place was at least three miles from any house, they could not conceive what could have taken him there. It was, probably, the influence of her will.

The power of *will* is a phenomenon that has been observed in all ages of the world, though of late years much less than at an earlier period ; and, as it was then more frequently exerted for evil than good, it was looked upon

as a branch of the art of black magic, whilst
the philosophy of it being unknown, the devil
was supposed to be the real agent, and the
witch, or wizard, only his instrument. The
profound belief in the existence of this art is
testified by the twelve tables of Rome, as well
as by the books of Moses, and those of Plato,
&c. It is extremely absurd to suppose that
all these statutes were erected to suppress a
crime which never existed; and, with regard
to these witches and wizards, we must remember, as Dr. Ennemoser justly remarks, that
the force of will has no relation to the strength
or weakness of the body; witness the extraordinary feats occasionally performed by feeble
persons under excitement, &c.; and, although
these witches and wizards were frequently
weak, decrepit people, they either believed in
their own arts, or else that they had a friend
or coadjutor in the devil, who was able and
willing to aid them. They, therefore, did not
doubt their own power, and they had the one
great requisite, *faith*. To *will and to believe*,
was the explanation given by the Marquis de
Puységur of the cures he performed; and this
unconsciously becomes the recipe of all
such men as Greatrix, the Shepherd of
Dresden, and many other wonder-workers,

and hence we see why it is usually the
humble, the simple and the child-like,
the solitary, the recluse, nay, the ignorant,
who exhibit traces of these occult faculties;
for he who cannot believe, cannot *will*, and
the scepticism of the intellect disables the
magician; and hence we see, also, wherefore,
in certain parts of the world and in certain
periods of its history, these powers and practices have prevailed. They were believed
in because they existed; and they existed
because they were believed in. There was a
continued interaction of cause and effect—of
faith and works. People who look superfi-
cially at these things, delight in saying that the
more the witches were persecuted the more
they abounded; and that when the persecution ceased we heard no more of them.
Naturally; the more they were persecuted the
more they believed in witchcraft and in themselves; when persecution ceased and men in
authority declared that there was no such
thing as witchcraft or witches, they lost their
faith; and with it, that little sovereignty over
nature that that faith had conquered.

Here we, also, see an explanation of the
power attributed to blessings and curses. The
Word of God is creative, and man is the child

of God, made in his image; who never out-
grows his childhood, and is often most a child
when he thinks himself the wisest, for " the
wisdom of this world," we cannot too often
repeat, " is foolishness before God"—and
being a child, his faculties are feeble in proportion ; but though limited in amount, they
are divine in kind, and are latent in all of us ;
still shooting up here and there, to amaze and
perplex the wise, and make merry the foolish,
who have nearly all alike forgotton their origin,
and disowned their birthright.

CHAPTER II.

TROUBLED SPIRITS.

A VERY curious circumstance, illustrative of
the power of will, was lately narrated to me
by a Greek gentleman, to whose uncle it occurred. His uncle, Mr. M., was, some years
ago travelling in Magnesia, with a friend, when
they arrived one evening at a caravanserai,
where they found themselves unprovided with
anything to eat. It was therefore agreed, that
one should go forth and endeavour to procure
food; and the friend offering to undertake the
office, Mr. M. stretched himself on the floor to

repose. Some time had elapsed and his friend had not yet returned, when his attention was attracted by a whispering in the room ; he looked up, but saw nobody, though still the whispering continued seeming to go round by the wall. At length it approached him ; but though he felt a burning sensation on his cheek, and heard the whispering distinctly, he could not catch the words. Presently he heard the footsteps of his friend, and thought he was returning ; but though they appeared to come quite close to him, and it was perfectly light, he still saw nobody ; then he felt a strange sensation—an irresistible impulsion to rise ; he felt himself *drawn up*, across the room, out of the door, down the stairs ; he must go, he could not help it, to the gate of the caravanserai, a little farther, and there he found the dead body of his friend, who had been suddenly assailed and cut down by robbers, unhappily too plenty in the neighbourhood at that period.

We here see the desire of the spirit to communicate his fate to the survivor ; the imperfection of the rapport or the receptivity, which prevented a more direct intercourse ; and the exertion of a magnetic influence, which Mr. M. could not resist, precisely similar to that of a living magnetiser over his patient.

There is a story extant in various English collections, the circumstances of which are said to have occurred about the middle of the last century, and which I shall here mention, on account of its similarity to the one that follows it.

Dr. Bretton, who was late in life appointed rector of Ludgate, lived previously in Herefordshire, where he married the daughter of Dr. Santer, a woman of great piety and virtue. This lady died: and one day as a former servant of hers, to whom she had been attached, and who had since married, was nursing her child in her own cottage, the door opened, and a lady entered so exactly resembling the late Mrs. Bretton in dress and appearance, that she exclaimed, " If my mistress were not dead, I should think you were she !" Whereupon, the apparition told her that she was so, and requested her to go with her, as she had business of importance to communicate. Alice objected, being very much frightened, and entreated her to address herself rather to Dr. Bretton, but Mrs. B. answered, *that she had endeavoured to do so, and had been several times in his room for that purpose, but he was still asleep, and she had no power to do more towards awakening him than once uncover his feet.* Alice then pleaded that she had nobody

to leave with her child, but Mrs. B. promising
that the child should sleep till her return, she
at length obeyed the summons, and, having
accompanied the apparition into a large field,
the latter bade her observe how much she
measured off with her feet, and, having
taken a considerable compass, she bade her go
and tell her brother that all that portion had
been wrongfully taken from the poor by their
father, and that he must restore it to them,
adding, that she was the more concerned about
it, since her name had been used in the trans-
action. Alice then asking how she should
satisfy the gentleman of the truth of her mission, Mrs. B. mentioned to her some circumstance known only to herself and this brother;
she then entered into much discourse with
the woman, and gave her a great deal of good
advice, remaining till hearing the sound of
horse-bells, she said, "Alice, I must be seen by
none but yourself," and then disappeared.
Whereupon, Alice proceeded to Dr. Bretton,
who admitted that he had actually heard some
one walking about his room, in a way he
could not account for. On mentioning the
thing to the brother, he laughed heartily, till
Alice communicated the secret which constituted her credentials, upon which he changed

his tone, and declared himself ready to make the required restitution.

Dr. Bretton seems to have made no secret of this story, but to have related it to various persons; and I think it is somewhat in its favour, that it exhibits a remarkable instance of the various degrees of receptivity of different individuals, where there was no suspicion of the cause, nor no attempt made to explain why Mrs. Bretton could not communicate her wishes to her husband as easily as to Alice. The promising that the child should sleep, was promising no more than many a magnetiser could fulfil. There are several curious stories, extant of lame and suffering persons, suddenly recovering, who attributed their restoration to the visit of an apparition which had stroked their limbs, &c.; and these are the more curious from the fact, that they occurred before Mesmer's time, when people in general knew nothing of vital magnetism. Dr. Binns quotes the case of a person named Jacob Olaffson, a resident in some small island, subject to Denmark, who after lying very ill for a fortnight, was found quite well, which he accounted for, by saying, that a person in shining clothes had come to him in the night, and stroked him with his hand, whereupon, he was presently

healed. But the stroking is not always necessary, since we know that the eye and the will
can produce the same effect.

The other case I alluded, as similiar to that of
Mrs. Bretton, occurred in Germany, and is related by Dr. Kerner.

The late Mr. L. St. ——, he says, quitted
this world with an excellent reputation, being
at the time superintendant of an institution
for the relief of the poor, in B—. His son in-
herited his property, and in acknowledgment
of the faithful services of his father's old house-
keeper, he took her into his family and established her in a country house, a few miles
from B—, which formed part of his inheritance.
She had been settled there but a short time,
when she was awakened in the night, she
knew not how, and saw a tall, haggard looking
man in her room, who was rendered visible to
her by a light that seemed to issue from himself. She drew the bed-clothes over her head;
but as this apparition appeared to her repeatedly, she became so much alarmed that
she mentioned it to her master, begging per-
mission to resign her situation. He however
laughed at her, told her it must be all imagination, and promised to sleep in the adjoining
apartment, in order that she might call him

whenever this terror seized her. He did so; but when the spectre retuined, she was so much oppressed with horror that she found it impossible to raise her voice. Her master then advised her to enquire the motive of its visits. This she did ; whereupon, it beckoned her to follow, which, after some struggles, she summoned resolution to do. It then led the way down some steps to a passage, where it pointed out to her a concealed closet, which it signified to her, by signs, she should open. She represented that she had no key, whereupon, it described to her, in sufficiently articulate words, where she would find one. She procured the key, and on opening the closet, found a small parcel, which the spirit desired her to remit to the governor of the institution for the poor, at B., with the injunction, that the contents should be applied to the benefit of the inmates ; this restitution being the only means, whereby he could obtain rest and peace in the other world. Having mentioned these circumstances to her master, who bade her do what she had been desired, she took the parcel to the governor and delivered it without communicating by what means it had come into her hands. Her name was entered in their books, and she was dismissed ; but after she was gone, they discovered, to their surprise, that the packet contained an order for thirty thousand florins, of which the late Mr. St. — had defrauded the institution and converted to his own use.

Mr. St. —, jun. was now called upon to pay the money, which he refusing to do, the affair was at length referred to the authorities and the house-keeper being arrested, he and she were confronted in the court, where she detailed the circumstances by which the parcel had come into her possession. Mr. St. —, denied the possibility of the thing, declaring the whole must be, for some purpose or other, an invention of her own. Suddenly, whilst making this defence he felt a blow upon his shoulder, which caused him to start and look round, and at the same moment the house-keeper exclaimed "See! there he stands now! there is the ghost!" None perceived the figure excepting the woman herself and Mr. St. —, but every body present, the minister included, heard the following words, " My son, repair the injustice I have committed, that I may be at peace!" The money was paid; and Mr. St. — was so much affected by this painful event, that he was seized with a severe illness, from which he with difficulty recovered.

Dr. Kerner says that these circumstances occurred in the year 1816, and created a considerable sensation at the time, though at the earnest request of the family of Mr. St. —, there was an attempt made to hush them up; adding, that in the month of October 1819, he was himself assured by a very respectable citizen of B., that it was universally known in the town, that the ghost of the late superintendant had appeared to the house-keeper, and pointed out to her where she would find the packet; that she had consulted the minister of her parish, who bade her deliver it as directed; that she had been subsequently arrested; and the affair brought before the authorities, where, whilst making his defence, Mr. St. —, had received a blow on the shoulder from an invisible hand; and that Mr. St. — was so much affected by these circumstances, which got abroad in spite of the efforts to suppress them, that he did not long survive the event.

Grose, the antiquary, makes himself very merry with the observation, that ghosts do not go about their business like other people; and

that in cases of murder, instead of going to
the nearest justice of peace or to the nearest
relation of the deceased, a ghost addresses itself to somebody who had nothing to do with
the matter, or hovers about the grave where
its body is deposited. " The same circuitous
mode is pursued," he says, " with respect to
redressing injured orphans or widows;
where it seems as if the shortest and most
certain way would be, to go and haunt the
person guilty of the injustice, till he were
terrified into restitution. We find the same
sort of strictures made on the story of the
ghost of Sir George Villiers, which instead of
going directly to his son, the Duke of Buck-
ingham to warn him of his danger, addressed
himself to an inferior person, whilst the
warning was after all inefficacious, as the
Duke would not take council; but surely
such strictures are as absurd as the conduct of
the ghost; at least I think there can be
nothing more absurd than pretending to prescribe laws to nature, and judging of what we
know so little about.

The proceedings of the ghost in the following case will doubtless be equally displeasing
to the critics. The account is extracted ver-
batim from a work published by the Banna-

tyne Club, and is entitled, "Authentic
Account of the Appearance of a Ghost in
Queen Ann's County, Maryland, United
States of North America, proved in the following remarkable trial, from attested notes,
taken in court at the time, by one of the
counsel."

It appears that Thomas Harris had made
some alteration in the disposal of his property,
immediately previous to his death; and that
the family disputed the will and raised up
difficulties likely to be injurious to his children.

" William Brigs said, that he was forty-
three years of age; that Thomas Harris died
in September, in the year 1790. In the March
following he was riding near the place where
Thomas Harris was buried, on a horse formerly belonging to Thomas Harris. After
crossing a small branch, his horse began to
walk on very fast. It was between the hours
of eight and nine o'clock in the morning. He
was alone: it was a clear day. He entered a
lane adjoining to the field where Thomas
Harris was buried. His horse suddenly
wheeled in a pannel of the fence, looked over
the fence into the field where Thomas Harris
was buried and neighed very loud. Witness

then saw Thomas Harris coming towards him, in the same apparel he had last seen him in his lifetime; he had on a sky blue coat. Just before he came to the fence, he varied to the right, and vanished; his horse immediately took the road. Thomas Harris came within two pannels of the fence to him; he did not see his features, nor speak to him. He was acquainted with Thomas Harris when a boy, and there had always been a great intimacy between them. He thinks the horse knew Thomas Harris, because of his neighing, pricking up his ears, and looking over the fence.

"About the first of June following he was ploughing in his own field, about three miles from where Thomas Harris was buried. About dusk Thomas Harris came alongside of him, and walked with him about two hundred yards. He was dressed as when first seen. He made a halt about two steps from him. J. Bailey who was ploughing along with him, came driving up, and he lost sight of the ghost. He was much alarmed: not a word was spoken. The young man Bailey did not see him; he did not tell Bailey of it. There was no motion of any particular part: he vanished. It preyed upon his mind so as to affect his health.

VOL. II.   E

He was with Thomas Harris when he died, but had no particular conversation with him. Some time after he was lying in bed, about eleven and twelve o'clock at night, he heard Thomas Harris groan, it was like the groan he gave a few minutes before he expired: Mrs. Brigs, his wife, heard the groan. She got up and searched the house: he did not, because he knew the groan to be from Thomas Harris. Some time after when in bed, and a great fire-light in the room, he saw a shadow on the wall at the same time he felt a great weight upon him. Some time after, when in bed and asleep, he felt a stroke between his eyes, which blackened them both: his wife was in bed with him, and two young men were in the room. The blow awaked him, and all in the room were asleep; is certain no person in the room struck him: the blow swelled his nose. About the middle of August he was alone, coming from Hickey Collins', after dark, about one hour in the night, Thomas Harris appeared, dressed as he had seen him when he was going down to the meeting house branch, three miles and a half from the grave-yard of Thomas Harris. It was star-light. He extended his arms over his shoulders. Does not know how long he remained in this situation. He was much alarmed. Thomas Harris disappeared. Nothing was said. He felt no weight on his shoulders. He went back to Colonel Linsi, and got a young man to go with him. After he got home he mentioned it to the young man. He had, before this, told James Harris he had seen his brother's ghost.

"In October, about twilight in the morning, he saw Thomas Harris about one hundred yards from the house of the witness; his head was leant to one side; same apparel as before ; his face was towards him; he walked fast and disappeared; there was nothing between them to obstruct the view ; he was about fifty yards from him, and alone ; he had no conception why Thomas Harris appeared to him. On the same day, about eight o'clock in the morning, he was handing up blades to John Bailey, who was stacking them; he saw Thomas Harris come along the garden-fence, dressed, as before he vanished, and always to the East : was within fifteen feet of him ; Bailey did not see him. An hour and a half afterwards, in the same place, he again appeared, coming as before; came up to the fence; leaned on it within ten feet of the witness, who called to Bailey to look there (pointing towards Thomas Harris). Bailey asked what was there ?

Don't you see Harris? Does not recollect
what Bailey said. Witness advanced towards
Harris. One or the other spoke as witness
got over the fence, on the same pannel that
Thomas Harris was leaning on. They
walked off together about five hundred yards,
a conversation took place as they walked;
he has not the conversation on his memory.
He could not understand Thomas Harris, his
voice was so low. He asked Thomas Harris
a question, and he forbid him. Witness then
asked, ' Why not go to your brother, instead
of me?' Thomas Harris said, ' Ask me no
questions.' Witness told him his will was
doubted. Thomas Harris told him to ask
his brother if he did not remember the conversation which passed between them on the
east side of the wheat stacks, the day he was
taken with his death sickness; that he then
declared that he wished all his property kept
together by James Harris, until his children
arrived at age, then the whole should be sold
and divided among his children; and, should
it be immediately sold, as expressed in his
will, that the property would be most wanting
to his children while minors, therefore he
had changed his will, and said that witness
should see him again. He then told witness

to turn, and disappeared. He did not speak
to him with the same voice as in his life-time.
He was not daunted while with Thomas
Harris, but much afterwards. Witness then
went to James Harris, and told him that he
had seen his brother three times that day.
Related the conversation he had with him.
Asked James Harris if he remembered the
conversation between him and his brother, at
the wheat-stack, he said he did, and told him
what had passed. Said he would fulfill his
brother's will. He was satisfied that witness
had seen his brother, for that no other person
knew the conversation. On the same evening,
returning home about an hour before sun-set,
Thomas Harris appeared to him, came alongside of him. Witness told him that his
brother said he would fulfill his will. No more
conversation on this subject. He disappeared.
He had further conversation with Thomas
Harris, but not on this subject. He was
always dressed in the same manner. He had
never related to any person the last conversation, and never would.

"Bailey, who was sworn in the cause, declared
that as he and Brigs were stacking blades, as
related by Brigs, he called to witness and

said, 'Look there! Do you not see Thomas
Harris?' Witness said, 'No.' Brigs got
over the fence, and walked some distance; appeared by his action to be in deep conversation
with some person. Witness saw no one.

"The counsel was extremely anxious to
hear from Mr. Briggs the whole of the conversation of the ghost, and on his cross-
examination took every means, without effect,
to obtain it. They represented to him, as a
religious man, he was bound to disclose the
whole truth. He appeared agitated when
applied to, declaring nothing short of life
should make him reveal the whole conversation, and claiming the protection of the
court, that he had declared all he knew relative
to the case.

"The Court overruled the question of the
counsel. Hon. James Tilgman, judge.

"His Excellency Robert Wright, late Governor of Maryland, and the Hon. Joseph H.
Nicholson, afterwards judge of one of the
courts in Maryland, were the counsel for the
plaintiff.

"John Scott and Richard T. Earle, Esqrs.,
were counsel for the defendant."

Here, as in the case of Colonel M., men-

tioned in a former chapter, and some others
I have met with, we find disclosures made
that were held sacred.

Dr. Kerner relates the following singular
story, which he declares himself to have received from the most satisfactory authority.
Agnes B., being at the time eighteen years
of age, was living as servant in a small inn at
Undenheim, her native place. The host and
hostess were quiet old people, who generally
went to bed about eight o'clock, whilst she
and the boy, the only other servant, were
expected to sit up till ten, when they had to
shut up the house and retire to bed also. One
evening, as the host was sitting on a bench
before the door, there came a beggar, requesting a night's lodging. The host, however, refused, and bade him seek what he
wanted in the village, whereon the man went
away.

At the usual hour the old people went to
bed, and the two servants, having closed the
shutters, and indulged in a little gossip with
the watchman, were about to follow their
example, when the beggar came round the
corner of the neighbouring street, and earnestly entreated them to give him a lodging
for the night, since he could find nobody that

would take him in. At first, the young people
refused, saying they dare not, without their
master's leave, but at length the entreaties of
the man prevailed, and they consented to let
him sleep in the barn, on condition that, when
they called him in the morning, he would
immediately depart. At three o'clock they
rose, and when the boy entered the barn, to his
dismay, he found that the old man had expired
in the night. They were now much perplexed
with the apprehension of their master's displeasure; so, after some consultation, they
agreed that the lad should convey the body
out of the barn, and lay it in a dry ditch
that was near at hand, where it would be
found by the labourers, and excite no question, as they would naturally conclude he had
laid himself down there to die.

This was done, the man was discovered and
buried, and they thought themselves well rid of
the whole affair; but, on the following night, the
girl was awakened by the beggar, whom she saw
standing at her bed-side. He looked at her,
and then quitted the room by the door. " Glad
was I," she says, " when the day broke, but I
was scarcely out of my room when the boy
came to me, trembling and pale, and before I
could say a word to him of what I had seen,

he told me that the beggar had been to his
room in the night, had looked at him, and
then gone away. He said he was dressed as
when we had seen him alive, only he looked
blacker, which I also had observed."

Still afraid of incurring blame, they told
nobody, although the apparition returned to
them every night, and although they found
removing to the other bed-chambers did not
relieve them from his visits. But the effects
of this persecution became so visible on both,
that much curiosity was awakened in the village with respect to the cause of the alteration
observed in them ; and, at length the boy's
mother went to the minister, requesting him
to interrogate her son, and endeavour to discover what was preying on his mind. To him
the boy disclosed their secret, and this minister,
who was a Protestant, having listened with attention to the story, advised him, when next
he went to Mayence, to market, to call on
Father Joseph, of the Franciscan Convent,
and relate the circumstance to him. This advice was followed, and Father Joseph, assuring
the lad that the ghost could do him no harm,
recommended him to ask him, in the name of
God, what he desired. The boy did so, whereupon the apparition answered, " Ye are chil-

dren of mercy, but I am a child of evil; in the barn, under the straw, you will find my money. Take it; it is yours." In the morning, the boy found the money accordingly in an old stocking, hid under the straw; but having a natural horror of it, they took it to their minister, who advised them to divide it into three parts: giving one to the Franciscan Convent, at Mayence, another to the Reformed Church, in the village, and the third to that to which they themselves belonged, which was of the Lutheran persuasion. This they did, and were no more troubled with the beggar. With respect to the minister who gave them this good advice, I can only say, all honour be to him! I wish there were many more such! The circumstance occurred in the year 1750, and is related by the daughter of Agnes B., who declared that she had frequently heard it from her mother.

The circumstance of this apparition looking darker than the man had done when alive, is significant of his condition; and confirms what I have said above, namely, that the moral state of the disembodied soul can no longer be concealed as it was in the flesh; but that as he is, he must necessarily appear.

There is an old saying, that we should never

lie down to rest at enmity with any human
being; and the story of the ghost of the
Princess Anna of Saxony, who apppeared to
Duke Christian of Saxe-Eisenburg, is strongly
confirmatory of the wisdom of this axiom.

Duke Christian was sitting one morning in
his study, when he was surprised by a knock
at his door—an unusual circumstance, since
the guards as well as the people in waiting
were always in the anti-room. He however
cried, " Come in!" when there entered to his
amazement, a lady in an ancient costume, who,
in answer to his enquiries, told him that she
was no evil spirit, and would do him no harm ;
but that she was one of his ancestors, and had
been the wife of Duke John Casimer, of Saxe-
Coburg. She then related that she and her
husband had not been on good terms at the
period of their deaths, and that although she
had sought a reconciliation, he had been in-
exorable; pursuing her with unmitigated
hatred, and injuring her by unjust suspicions ;
and that consequently although *she* was happy,
he was still wandering in cold and darkness,
betwixt time and eternity. She had, however,
long known that one of their descendants was
destined to effect this reconciliation for them,
and they were rejoiced to find the time for

it had at length arrived. She then gave
the Duke eight days to consider if he were
willing to perform this good office, and disappeared; wherenpon he consulted a clergyman, in whom he had great confidence, who
after finding the ghosts communications verified, by a reference to the annals of the family,
advised him to comply with her request.

As the Duke had yet some difficulty in
believing it was really a ghost he had seen,
he took care to have his door well watched;
she, however, entered at the appointed time,
unseen by the attendants; and, having received the Duke's promise, she told him that
she would return with her husband on the
following night; for that, though she could
come by day, he could not; that then,
having heard the circumstances, the Duke
must arbitrate between them, and then unite
their hands, and bless them. The door was
still watched, but nevertheless the apparitions
both came, the Duke Casimer in full royal
costume, but of a livid paleness; and when
the wife had told her story, he told his. Duke
Christian decided for the lady, in which judgment Duke Casimer fully acquiesced. Christian then took the ice-cold hand of Casimer
and laid it in that of his wife, which felt of a

natural heat. They then prayed and sang together, and the apparitions disappeared, having foretold that Duke Christian would ere long be with them. The family records showed that these people had lived about one hundred years before Duke Christian's time, who himself died in the year 1707, two years after these visits of his ancestors. He desired to be buried in quick lime, it is supposed from an idea that it might prevent his ghost walking the earth.

The costume in which they appeared was precisely that they had worn when alive; as was ascertained by a reference to their portraits.

The expression, that her husband was *wandering in cold and darkness, betwixt time and eternity*, are here, very worthy of observation; as are the circumstances, that his hand was cold, whilst hers was warm; and also, the greater privilege she seemed to enjoy. The hands of the unhappy spirits appear, I think, invariably to communicate a sensation of cold.

I have heard of three instances, of persons now alive, who declare that they hold continual intercourse with their deceased partners. One of these, is a naval officer, whom the author of the book lately published, called "The

Unseen World," appears to be acquainted
with. The second is a professor, in a college
in America, a man of eminence and learning,
and full of activity and energy—yet, he assured a friend of mine, that he receives constant visits from his departed wife, which afford
him great satisfaction. The third example, is
a lady in this country. She is united to a second
husband, has been extremely happy in
both marriages ; and declares that she receives
frequent visits from her first. Oberlin, the
good pastor of *Ban de la Roche,* asserted the
same thing of himself. His wife came to him
frequently after her death ; was seen by the
rest of his household, as well as himself ; and
warned him, beforehand of many events that
occurred.

Mrs. Mathews relates in the memoirs of
her husband, that he was one night in bed
and unable to sleep, from the excitement that
continues some time after acting ; when hearing a rustling by the side of the bed, he
looked out, and saw his first wife, who was
then dead, standing by the bed-side, dressed
as when alive. She smiled and bent forward
as if to take his hand; but in his alarm, he
threw himself out on the floor to avoid the
contact, and was found by the landlord in a

fit. On the same night and at the same hour, the present Mrs. Mathews who was far away from him, received a similar visit from her predecessor, whom she had known when alive. She was quite awake ; and in her terror seized the bell rope to summon assistance, which gave way, and she fell with it in her hand, to the ground.

Professor Barthe, who visited Oberlin in 1824, says, that whilst he spoke of his intercourse with the spiritual world as familiarly as of the daily visits of his parishioners, he was at the same time, perfectly free from fanaticism, and eagerly alive to all the concerns of this earthly existence. He asserted, what I find many somnambules and deceased persons also assert, that everything on earth is but a copy, of which the antitype is to be found in the other.

He said to his visitor, that he might as well attempt to persuade him that that was not a table before them, as that he did not hold communication with the other world. " I give you credit for being honest, when you assure me that you never saw anything of the kind," said he ; " give me the same credit when I assure you that I do."

With respect to the faculty of ghost-seeing,

he said, it depended on several circumstances, external and internal. People who live in the bustle and glare of the world seldom see them, whilst those who live in still, solitary, thinly inhabited places, like the mountainous districts of various countries, do. So if I go into the forest by night, I see the phosphoric light of a piece of rotten wood; but if I go by day I cannot see it; yet it is still there. Again, there must be a rapport. A tender mother is awakened by the faintest cry of her infant, whilst the maid slumbers on, and never hears it; and if I thrust a needle amongst a parcel of wood-shavings, and hold a magnet over them, the needle is stirred, whilst the shavings are quite unmoved. There must be a particular aptitude: what it consists in, I do not know; for of my people, many of whom are ghost-seers, some are weak and sickly, others vigorous and strong. Here are several pieces of flint: I can see no difference in them; yet some have so much iron in them that they easily become magnetic; others have little or none. So it is with the faculty of ghost-seeing. People may laugh as they will, but the thing is a fact, nevertheless.

The visits of his wife continued for nine years after her death, and then ceased.

At length she sent him a message, through another deceased person, to say that she was now elevated to a higher state, and could therefore no longer revisit the earth.

Never was there a purer spirit, nor a more beloved human being, than Oberlin. When first he was appointed to the cure of Ban de la Roche, and found his people talking so familiarly of the re-appearance of the dead, he reproved them and preached against the superstition; nor was he convinced, till after the death of his wife. She had however previously received a visit from her deceased sister, the wife of Professor Oberlin, of Strasburg, who had warned her of her approaching death, for which she immediately set about preparing, making extra clothing for her children, and even laying in provisions for the funeral feast. She then took leave of her husband and family, and went quietly to bed. On the following morning she died; and Oberlin never heard of the warning she had received, till she disclosed it to him in her spectral visitations.

In narrating the following story, I am not permitted to give the names of the place or parties, nor the number of the regiment, with all of which however I am acquainted. The account was taken down by one of the officers,

with whose family I am also acquainted; and
the circumstance occurred within the last
ten years.

" About the month of August," says Captain
E., " my attention was requested by the school-
master-sergeant, a man of considerable worth,
and highly esteemed by the whole corps, to an
event which had occurred in the garrison hospital. Having heard his recital, which, from
the serious earnestness with which he made it,
challenged attention, I resolved to investigate
the matter; and having communicated the
circumstances to a friend, we both repaired to
the hospital for the purpose of enquiry.

"There were two patients to be examined—
both men of good character, and neither of
them suffering from any disorder affecting the
brain; the one was under treatment for consumptive symptoms, and the other for an
ulcerated leg ; and they were both in the prime
of life.

" Having received a confirmation of the
schoolmaster's statement from the hospital-
sergeant, also a very respectable and trustworthy man, I sent for the patient principally
concerned, and desired him to state what he
had seen and heard, warning him, at the same
time, that it was my intention to take down his

deposition, and that it behoved him to be very careful, as, possibly, serious steps might be taken for the purpose of discovering whether an imposition had been practised in the wards of the hospital—a crime for which, he was well aware, a very severe penalty would be inflicted. He then proceeded to relate the circumstances, which I took down in the presence of Mr. B., and the hospital-sergeant, as follows :—

"'It was last Tuesday night, somewhere between eleven and twelve, when all of us were in bed, and all lights out except the rush-light that was allowed for the man with the fever, when I was awoke by feeling a weight upon my feet, and at the same moment, as I was drawing up my legs, Private W., who lies in the cot apposite mine, called out, "I say, Q. there's somebody sitting upon your legs!" and as I looked to the bottom of my bed, I saw some one get up from it, and then come round and stand over me, in the passage betwixt my cot and the next. I felt somewhat alarmed ; for the last few nights the ward had been disturbed by sounds as of a heavy foot walking up and down; and as nobody could be seen, it was beginning to be supposed amongst us that it was haunted, and fancying this that came up to my bed's head might be the ghost,

I called out, " Who are you ? and what do you
want ? "

" ' The figure then leaning, with one hand
on the wall, over my head, and stooping down,
said, in my ear, " I am Mrs. M. ;" and I could
then distinguish that she was dressd in a
flannel gown, edged with black ribbon, exactly
similar to a set of grave clothes in which I had
assisted to clothe her corpse, when her death
took place a year previously.

" ' The voice however was not like Mrs. M.'s,
nor like anybody elses, yet it was very distinct,
and seemed somehow to sing through my
head. I could see nothing of a face beyond a
darkish colour about the head, and it appeared
to me that I could see through her body against
the window-glasses.

" ' Although I felt very uncomfortable, I
asked her what she wanted. She replied, " I
am Mrs. M., and I wish you to write to him
that was my husband, and tell him........"

" ' I am not, sir,' said Corporal Q., ' at liberty
to mention to anybody what she told me,
except to her husband. He is at the depot in
Ireland, and I have written and told him. She
made me promise not to tell any one else.
After I had promised secrecy, she told me
something of a matter, that convinced me I

was talking to a spirit; for it related to what
only I and Mrs. M. knew, and no one living
could know anything whatever of the matter;
and if I was now speaking my last words on
earth, I say solemnly that it was Mrs. M.'s
spirit that spoke to me then, and no one else.
After promising that if I complied with her
request, she would not trouble me or the ward
again, she went from my bed towards the fire-
place, and with her hands she kept feeling
about the wall over the mantel-piece. After
awhile, she came towards me again; and whilst
my eyes were upon her, she somehow disappeared from my sight altogether, and I was
left alone.

"'It was then that I felt faint like, and a
cold sweat broke out over me; but I did not
faint, and after a time I got better, and gradually I went off to sleep.

"'The men in the ward said, next day, that
Mrs. M, had come to speak to me about purgatory, because she had been a Roman Catholic, and we had often had arguments on
religion: but what she told me had no reference to such subjects, but to a matter only she
and I knew of.'

"After closely cross-questioning Corporal
Q., and endeavouring, without success, to

reason him out of his belief in the ghostly
character of his visitor; I read over to him
what I had written, and then, dismissing him,
sent for the other patient.

"After cautioning him, as I had done the
first, I proceeded to take down his statement,
which was made with every appearance of good
faith and sincerity.

"'I was lying awake,' said he, 'last Tuesday
night, when I saw some one sitting on Corporal
Q.'s bed. There was so little light in the
ward that I could not make out who it was,
and the figure looked so strange that I got
alarmed, and felt quite sick. I called out to
Corporal Q. that there was somebody sitting
upon his bed, and then the figure got up; and
as I did not know but it might be coming to
me, I got so much alarmed, that being but
weakly (this was the consumptive man) I fell
back, and I believe I fainted away. When I
got round again, I saw the figure standing,
and apparently talking to the Corporal, placing
one hand against the wall and stooping down.
I could not however hear any voice; and being
still much alarmed, I put my head under the
clothes for a considerable time. When I
looked up again I could only see Corporal Q.,
sitting up in bed alone, and he said he had

seen a ghost; and I told him I had also seen
it. After a time, he got up and gave me a
drink of water, for I was very faint. Some
of the other patients being disturbed by our
talking, they bid us be quiet, and after some
time I got to sleep. The ward has not been
disturbed since.'

" The man was then cross-questioned; but
his testimony remaining quite unshaken, he
was dismissed, and the hospital-sergeant was
interrogated, with regard to the possibility of
a trick having been practised. He asserted,
however, that this was impossible; and, certainly, from my own knowledge of the hospital
regulations, and the habits of the patients, I
should say that a practical joke of this nature
was too serious a thing to have been attempted
by anybody, especially as there were patients
in the ward very ill at the time, and one very
near his end. The punishment would have
been extremely severe, and discovery almost
certain, since everybody would have been
adverse to the delinquent.

" The investigation that ensued was a very
brief one, it being found that there was nothing more to be elicited; and the affair terminated with the supposition that the two
men had been dreaming. Nevertheless, six

months afterwards, on being interrogated, their evidence and their conviction were as clear as at first, and they declared themselves ready at any time to repeat their statement upon oath."

Supposing this case to be as the men believed it, there are several things worthy of observation. In the first place, the ghost is guilty of that inconsistency so offensive to Francis Grose and many others. Instead of telling her secret to her husband, she commissions the Corporal to tell it him, and it is not till a year after her departure from this life that she does even that; and she is heard in the ward two or three nights before she is visible. We are therefore constrained to suppose that like Mrs. Bretton, she could not communicate with her husband, and that till that Tuesday night, the necessary conditions for attaining her object, as regarded the Corporal, were wanting. It is also remarkable, that although the latter heard her speak distinctly, and spoke to her, the other man heard no voice; which renders it probable, that she had at length been able to produce that impression upon him, which a magnetiser does on his somnambule, enabling each to understand the other by a transference of thought, which

was undistinguishable to the Corporal from
speaking, as it is frequently to the somnambule.
The imitating the actions of life by leaning
against the wall and feeling about the mantel-
piece, are very unlike what a person would
have done, who was endeavouring to impose
on the man; and equally unlike what they
would have repoited, had the thing been an
invention of their own.

Amongst the established jests on the subject
of ghosts, their sudden vanishing, is a very
fruitful one; but, I think, if we examine
this question, we shall find, that there is nothing comical in the matter, except the ignorance or want of reflection of the jesters.

In the first place, as I have before observed,
a spirit must be where its thoughts and affections are, for they are itself—*our* spirits are
where our thoughts and affections are, although our solid bodies remain stationary; and
no one will suppose, that walls or doors, or
material obstacles of any kind, could exclude
a spirit, any more than they can exclude our
thoughts.

But then, there is the visible body of the
spirit—what is that? and how does it retain
its shape! For we know, that there is a law
discovered by Dalton, that two masses of

gaseous matter cannot remain in contact, but
they will immediately proceed to diffuse themselves into one another; and accordingly, it
may be advanced, that a gaseous corporeity in
the atmosphere, is an impossibility, because it
could not retain its form, but would inevitably
be dissolved away, and blend with the surrounding air. But precisely the same objection
might be made by a chemist to the possibility
of our fleshly bodies retaining their integrity
and compactness : for the human body, taken
as a whole, is knwn to be an impossible
chemical compound, except for the vitality
which upholds it; and no sooner is life withdrawn from it, than it crumbles into putrescence;
and it is undeniable, that the æriform body
would be an impossible mechanical phenomenon, but for the vitality which, we are
entitled to suppose, may uphold it. But, just
as the state or condition of organisation pro-
tects the fleshly body from the natural re-actions
which would destroy it, so may an analogous
condition of organisation protect a spiritual
ethereal body from the destructive influence of
the mutual inter-diffusion of gases.

Thus, supposing this æriform body to be a
permanent appurtenance of the spirit, we see
how it may subsist and retain its integrity, and

it would be as reasonable to hope to exclude
the electric fluid by walls or doors as to exclude
by them this subtle, fluent form. If, on the
contrary, the shape be only one constructed
out of the atmosphere, by an act of will, the
same act of will, which is a vital force, will
preserve it entire, till the will being withdrawn,
it dissolves away. In either case, the moment
the will or thought of the spirit is elsewhere, it
is gone—it has vanished.

For those who prefer the other hypothesis,
namely, that there is no outstanding shape at
all, but that the will of the spirit, acting on
the constructive imagination of the seer, enables him to conceive the form, as the spirit
itself conceives of it, there can be no difficulty
in understanding, that the becoming invisible
will depend merely on a similar act of will.

CHAPTER III.

HAUNTED HOUSES.

EVERYBODY has heard of haunted houses; and there is no country, and scarcely any place, in which something of the sort is not known or talked off; and I suppose there in no one who, in the course of their travels, has not seen very respectable, good-looking houses shut up and uninhabited, because they had this evil reputation assigned to them. I have seen several such, for my own part; and it is remarkable that this *mala fama* does not always, by any means, attach itself to buildings one would imagine most obnoxious to such a suspicion.

For example, I never heard of a ghost being
seen or heard in Haddon Hall, the most ghostly
of houses; nor in many other antique, mysterious looking buildings, where one might expect them, whilst sometimes a house of a very
prosaic aspect remains uninhabited, and is ultimately allowed to fall to ruin for no other
reason, we are told, than that nobody can live
in it. I remember, in my childhood, such a
house in Kent—I think it was on the road betwixt Maidstone and Tunbridge—which had this
reputation. There was nothing dismal about
it; it was neither large nor old; and it stood
on the borders of a well frequented road; yet,
I was assured it had stood empty for years;
and as long as I lived in that part of the country
it never had an inhabitant, and I believe was
finally pulled down; and all for no other reason
than that it was haunted, and nobody could
live in it. I have frequently heard of people,
whilst travelling on the continent, getting into
houses at a rent so low as to surprise them, and
I have moreover frequently heard of very
strange things occurring whilst they were
there. I remember, for instance, a family of
the name of S. S., who obtained a very handsome house at a most agreeably cheap rate,
somewhere on the coast of Italy—I think it

was at Mola de Gaeta.   They lived very comfortably in it, till one day, whilst Mrs. S. S.
was sitting in the drawing-room, which opened
into a balcony overhanging the sea, she saw a
lady dressed in white pass along before the
windows, which were all closed.   Concluding
it was one of her daughters, who had been accidentally shut out, she arose and opened the
window to allow her to enter, but on looking
out, to her amazement, there was nobody there,
although there was no possible escape from the
balcony unless by jumping into the sea.   On
mentioning this circumstance to somebody in
the neighbourhood, they were told "that that
was the reason they had the house so cheap ;
nobody liked to live in it."

I have heard of several houses, even in
populous cities, to which some strange circumstance of this sort is attached—some in London even, and some in this city and neighbourhood ; and what is more, unaccountable things
actually do happen to those who inhabit them.
Doors are strangely opened and shut, a rustling
of silk, and sometimes a whispering, and frequently footsteps are heard.   There is a house
in Ayrshire, to which this sort of thing has
been attached for years, insomuch that it was
finally abandoned to an old man and woman,

who said that they were so used to it that they did not mind it. A distinguished authoress told me, that some time ago she passed a night at the house of an acquaintance, in one of the midland counties of England. She and her sister occupied the same room, and in the night they heard some one ascending the stairs; the foot came distinctly to the door, then turned away, ascended the next flight, and they heard it over head. In the morning, on being asked if they had slept well, they mentioned this circumstance. " That is what everybody hears who sleeps in that room," said the lady of the house. " Many a time I have, when sleeping there, drawn up the night-bolt, persuaded that the nurse was bringing the baby to me; but there was nobody to be seen. We have taken every pains to discover what it is, but in vain; and are now so used to it, that we have ceased to care about the matter."

I know of two or three other houses in this city, and one in the neighbourhood, in which circumstances of this nature are transpiring, or have transpired very lately; but people hush them up, from the fear of being laughed at, and also from an apprehension of injuring the character of a house; on which account, I do not dwell on the particulars; but there was

some time since a *fama* of this kind attached
to a house in St. J— Street, some of the details
of which became very public. It had stood
empty a long time, in consequence of the annoyances to which the inhabitants had been
subjected. There was one room particularly
which noboby could occupy without disturbance. On one occasion, a youth who had been
abroad a considerable time, either in the army
or navy, was put there to sleep on his arrival,
since knowing nothing of these reports, it was
hoped his rest might not be interrupted. In
the morning, however, he complained of the
dreadful time he had had with people looking
in at him between the curtains of his bed, all
night, avowing his resolution to terminate
his visit that same day, as he would not sleep
there any more. After this period, the house
stood empty again for a considerable time, but
was at length taken, and workmen sent in to
repair it. One day, when the men were away
at their dinner, the master tradesman took the
key, and went to inspect progress, and having
examined the lower rooms, he was ascending
the stairs, when he heard a man's foot behind
him. He looked round, but there was nobody
there, and he moved on again ; still there was
somebody following, and he stopped and looked

over the rails; but there was no one to be seen. So, though feeling rather queer, he advanced into the drawing-room, where a fire had been lighted, and wishing to combat the uncomfortable sensation that was creeping over him, he took hold of a chair, and drawing it resolutely along the floor, he slammed it down upon the hearth with some force, and seated himself in it; when, to his amazement, the action, in all its particulars of sound, was immediately repeated by his unseen companion, who seemed to seat himself beside him on a chair as invisible as himself. Horror-struck, the worthy builder started up and rushed out of the house.

There is a house in S— Street, in London, which, having stood empty a good while, was at length taken by Lord B. The family were annoyed by several unpleasant occurrences, and by the sound of footsteps, which were often audible, especially in Lady B.'s bed-room, who though she could not see the form, was occasionally conscious of its immediate proximity.

Some time since, a gentleman having established himself in a lodging in London, felt, the first night he slept there, that the clothes were being dragged off his bed. He fancied he had done it himself in his sleep, and pulled them on again; but it happens repeatedly; he

gets out of bed each time, can find nobody—
no string—no possible explanation, nor can
obtain any from the people of the house, who
only seem distressed and annoyed. On mentioning it to some one in the neighbourhood,
he is informed that the same thing has occurred
to several preceding occupants of the lodging ;
which, of course, he left.

The circumstances that happened at New
House, in Hampshire, as detailed by Mr.
Barham, in the third volume of the "Ingoldsby
Legends," are known to be perfectly authentic,
as are the following, the account of which I
have received from a highly respectable servant, residing in a family, with whom I am
well acquainted :—She informs me, that she
was not very long since living with a Colonel
and Mrs. W., who, being at Carlisle engaged
a furnished house, which they obtained at an
exceedingly cheap rate, because nobody liked
to live in it. This family, however, met with
no annoyance, and attached no importance to
the rumour which had kept the house empty.
There were, however, two rooms in it wholly
unfurnished, and as the house was large, they
were dispensed with, till the recurrence of the
race week, when, expecting company, these
two rooms were temporarily fitted up for the

use of the nurses and children. There were
heavy Venetian blinds to the windows, and in
the middle of the night, the person who related the circumstance to me, was awaked by
the distinct sound of these blinds being pulled
up and down with violence, perhaps as many
as twenty times. The fire had fallen low, and
she could not see whether they were actually
moved, or not, but lay trembling in indescribable terror. Presently, feet were heard in
the room, and a stamping as if several men
were moving about without stockings. Whilst
lying in this state of agony, she was comforted
by hearing the voice of a nurse, who slept in
another bed in the same chamber, exclaiming,
" The Lord have mercy upon us!" This second
woman then asked the first, if she had courage
to get out of bed and stir up the fire, so that
they might be able to see; which, by a great
effort, she did; the chimney being near her
bed. There was, however, nothing to be discovered; everything being precisely as when
they went to bed. On another occasion, when
they were sitting, in the evening at work, they
distinctly heard some one counting money and
the chink of the pieces as they were laid down.
The sound proceeded from the inner room of
the two; but there was nobody there. This

family left the house, and though a large and commodious one, she understood it remained unoccupied, as before.

A respectable citizen of Edinburgh, not long ago, went to America to visit his son, who had married and settled there. The morning after his arrival, he declared his determination to return immediately to Philadelphia, from which the house was at a considerable distance; and on being interrogated as to the cause of this sudden departure, he said that in the previous night he had heard a man walking about his room, who had approached the bed, drawn back the curtains, and bent over him. Thinking it was somebody who had concealed himself there with ill intentions, he had struck out violently at the figure; when, to his horror, his arm passed unimpeded through it.

Other extraordinary things happened in that house, which had the reputation of being haunted, although the son had not believed it, and had therefore not mentioned the report to the father. One day, the children said they had been running after "such a queer thing in the cellar; it was like a goat, and not like a goat; but it seemed to be like a shadow,"

A few years ago, some friends of mine were

taking a house in this city, when the servants
of the people who were leaving, advised them
not to have anything to do with it; for that
there was a ghost in it that screamed dreadfully, and that they never could keep a stitch
of clothes on them at night; the bed-coverings were always pulled off.   My friends
laughed heartily, and took the house; but the
cries and groans all over it were so frequent
that they at length got quite used to them.
It is to be observed that the house was a *flat*
or *floor*, shut in; so that there could be no
draughts of air nor access for tricks.   Besides,
it was a woman's voice, sometimes close to
their ears, sometimes in a closet, sometimes
behind their beds—in short, in all directions.
Everybody heard it that went to the house.

The tenant that succeeded them, however,
has never been troubled with it.

The story of the Brown Lady at the Marquis of T.'s, in Norfolk, is known to many.
The Hon. H. W. told me that a friend of his,
whilst staying there, had often seen her, and
had one day enquired of his host, " Who was
the lady in brown that he had met frequently
on the stairs ?" Two gentlemen, whose names
were mentioned to me, resolved to watch for
her and intercept her. They at length saw

VOL. II.   H

her, but she eluded them by turning down a staircase, and when they looked over she had disappeared. Many persons have seen her.

There is a Scotch family of distinction, who, I am told, are accompanied by an unseen attendant, whom they call " Spinning Jenny." She is heard spinning in their house in the country, and when they come into town, she spins here; servants and all hear the sound of her wheel. I believe she accompanies them no further than to their own residences, not to those of other people. Jenny is supposed to be a former housemaid of the family, who was a great spinner, and they are so accustomed to her presence as to feel it no annoyance.

The following very singular circumstance was related to me by the daughter of the celebrated Mrs. S :—Mrs. S. and her husband were travelling into Wales, and had occasion to stop on their way, some days, at Oswestry. There they established themselves in a lodging, to reach the door of which they had to go down a sort of close, or passage. The only inhabitants of the house were the mistress, a very handsome woman, and two maids. Mr. and Mrs. S., however, very soon had occason to complain of the neglected state of the rooms, which were apparently never cleaned or dusted; though, strange to say, to judge by their own ears, the servants were doing nothing else all night, their sleep being constantly disturbed by the noise of rubbing, sweeping, and the moving of furniture. When they complained to these servants of the noise in the night, and the dirt of the rooms, they answered that the noise was not made by them, and that it was impossible for them to do their work, exhausted as they were by sitting up all night with their mistress, who could not bear to be alone when she was in bed. Mr. and Mrs. S. afterwards discovered that she had her room lighted up every night; and one day, as they were returning from a walk, and she happened to be going down the close before them, they heard her saying, as she turned her head sharply from side to side, "Are you there again? What, the devil! Go away, I tell you! &c. &c." On applying to the neighbours for an explanation of these mysteries, the good people only shook their heads, and gave mysterious answers. Mr. and Mrs. S. afterwards learnt that she was believed to have murdered a girl who formerly lived in her service.

There is nothing in the conduct of this un-

happy woman which may not be perfectly
well accounted for, by the supposition of a
guilty conscience; but the noises heard by
Mr. and Mrs. S. at night, are curiously in accordance with a variety of similar stories,
wherein this strange visionary repetition of
the trivial actions of daily life, or of some
particular incident, have been observed. The
affair of Lord St. Vincent's was of this nature ;
and there is somewhere extant, an account of
the ghost of Peter the Great of Russia having
appeared to Doctor Doppelio, complaining to
him of the sufferings he endured from having
to act over again his former cruelties; a circumstance which exhibits a remarkable coin-
cidence with the Glasgow dream, mentioned
in a preceding chapter. We must, of course,
attach a symbolical meaning to these phenomena, and conclude that these reactings are
somewhat of the nature of our dreams.

Certainly, there would need no stronger
motive to induce us to spend the period allotted
to us on earth, in those pure and innocent
pleasures and occupations, which never weary
or sicken the soul, than the belief that such a
future awaits us !

A family in one of the English counties, was
a few years ago terribly troubled by an unseen

inmate, who chiefly seemed to inhabit a large
cellar, into which there was no entrance except
the door, which was kept locked. Here there
would be a loud knocking—sometimes a voice
crying—heavy feet walking, &c. &c. At first,
the old trustworthy butler would summon his
accolytes, and descend, armed with sword and
blunderbuss; but no one was to be seen.
They could often hear the feet following them
up stairs from this cellar; and once, when the
family had determined to watch, they found
themselves accompanied up stairs not only by
the sound of the feet, but by a *visible* shadowy
companion! They rushed up, flew to their
chamber, and shut the door, when instantly
they felt and saw the handle turned in their
hand by a hand outside. Windows and doors
were opened in spite of locks and keys; but
notwithstanding the most persevering investigations, the only clue to the mystery was the
appearance of that spectral figure.

The knockings and sounds of people at
work, asserted to be heard in mines, is a fact
maintained by many very sensible men, over-
seers, and superintendents, &c. as well as by
the workmen themselves; and there is a strong
persuasion, I know, amongst the miners of
Cornwall and those of Mendip, that these

visionary workmen are sometimes heard
amongst them ; on which occasions the horses
evince their apprehensions by trembling
and sweating; but as I have no means of
verifying these reports, I do not dwell upon
them further.

When the mother of George Canning, then
Mrs. Hunn, was an actress in the provinces, she
went, amongst other places, to Plymouth,
having previously requested her friend, Mr.
Bernard, of the theatre, to procure her a
lodging. On her arrival, Mr. B. told her that
if she was not afraid of a ghost, she might
have a comfortable residence at a very low
rate, " For there is," said he, " a house belonging to our carpenter, that is reported to be
haunted, and nobody will live in it. If you
like to have it, you may, and for nothing, I
believe, for he is so anxious to get a tenant;
only you must not let it be known that you do
not pay rent for it."

Mrs. Hunn, alluding to the theatrical apparitions, said, it would not be the first time she
had had to do with a ghost, and that she was
very willing to encounter this one ; so she
had her luggage taken to the house in question,
and the bed prepared. At her usual hour, she
sent her maid and her children to bed, and,

curious to see if there was any foundation for
the rumour she had heard, she seated herself
with a couple of candles and a book, to watch
the event. Beneath the room she occupied
was the carpenter's workshop, which had two
doors; the one which opened into the street
was barred and bolted within; the other, a
smaller one, opening into the passage, was
only on the latch; and the house was, of
course, closed for the night. She had read
something more than half an hour, when she
perceived a noise issuing from this lower
apartment, which sounded very much like the
sawing of wood; presently, other such noises
as usually proceed from a carpenter's workshop were added, till, by and by, there was a
regular concert of knocking and hammering,
and sawing and planing, &c.; the whole
sounding like half a dozen busy men in full
employment. Being a woman of considerable
courage, Mrs. Hunn resolved, if possible, to
penetrate the mystery; so taking off her shoes,
that her approach might not be heard, with
her candle in her hand, she very softly opened
her door and descended the stairs, the noise
continuing as loud as ever and evidently proceeding from the workshop, till she opened the
door, when instantly all was silent—all was

still—not a mouse was stirring; and the tools
and the wood, and everything else, lay as they
had been left by the workmen when they went
away. Having examined every part of the
place, and satisfied herself that there was nobody there, and that nobody could get into it,
Mrs. Hunn ascended to her room again, beginning almost to doubt her own senses, and
question with herself whether she had really
heard the noise or not, when it re-commenced
and continued, without intermission, for about
half an hour. She however went to bed, and
the next day told nobody what had occurred,
having determined to watch another night before mentioning the affair to any one. As,
however, this strange scene was acted over
again, without her being able to discover the
cause of it, she now mentioned the circumstance to the owner of the house and to her
friend Bernard; and the former, who would
not believe it, agreed to watch with her, which
he did. The noise began as before, and he
was so horror-struck, that instead of entering
the workshop, as she wished him to do, he
rushed into the street. Mrs. Hunn continued
to inhabit the house the whole summer, and
when referring afterwards to the adventure,
she observed, that use was second nature; and

that she was sure if any night these ghostly carpenters had not pursued their visionary labours, she should have been quite frightened, lest they should pay her a visit up stairs.

From many recorded cases, I find the vulgar belief, that buried money is frequently the cause of these disturbances, is strongly borne out by facts. This certainly does seem to us very strange ; and can only be explained by the hypothesis suggested, that the soul awakes in the other world in exactly the same state in which it quitted this.

In the above mentioned instances, of what are called *haunted houses*, there is generally nothing seen, but those are equally abundant, where the ghostly visitor is visible.

Two young ladies were passing the night in a house in the north, when the youngest, then a child, awoke and saw an old man, in a Kilmarnock night cap, walking about their bed room. She said, when telling the story in after life, that she was not the least frightened, she was only surprised ! but she found that her sister, who was several years older than herself, was in a state of great terror. He continued some time moving about, and at last went to a chest of drawers, where there lay a parcel of buttons, belonging to a travelling

tailor, who had been at work in the house.
Whether the old man threw them down, or
not, she could not say, but, just then, they all
fell rattling off the drawers to the floor, whereupon, he disappeared.   The next morning,
when they mentioned the circumstance, she
observed, that the family looked at each other
in a significant manner; but it was not till she
was older, she learnt, that the house was said
to be haunted by this old man,   " It never occurred to me," she said, " that it was a ghost
—who could have thought of a ghost in a
Kilmarnock night cap."

At the Leipsick fair, lodgings are often
very scarce, and on one occasion, a stranger
who had arrived late in the evening, had some
difficulty in finding a bed.   At length, he found
a vacant chamber in the house of a citizen;
it was one they made no nse of, but they said
he was welcome to it ; and weary and sleepy,
he gladly accepted the offer.   Fatigued as he
was, however, he was disturbed by some unaccountable noises, of which he complained
to his hosts in the morning.   They pacified
him by some excuses, but the next night, not
long after he had gone to bed, he came down
stairs in great haste, with his portmanteau on
his shoulder, declaring he would not stay

there another hour, for the world; for that a
lady in a strange old fashioned dress had come
into the room with a dagger in her hand, and
made threatening gestures at him. He accordingly went away, and the room was shut
up again; but some time afterwards, a servant
girl in the family of this citizen, being taken
ill, they were obliged to put her into that
room, in order to separate her from the rest of
the family. Here she recovered her health
rapidly, and as she had never complained of
any annoyance, she was asked, when she
was quite well, whether anything particular
had happened whilst she inhabited that
chamber, "Oh yes," she answered; "every
night there came a strange lady into the room,
who sat herself on the bed and stroked me
with her hand, and I believe it is to her I owe
my speedy recovery; but I could never get
her to speak to me—she only sighs and
weeps."

Not very long since, a gentleman set out,
one fine midsummer's evening, when it is light
all night in Scotland, to walk from Montrose
to Brechin. As he approached a place called
Dunn, he observed a lady walking on before,
which from the lateness of the hour, somewhat surprised him. Some time afterwards,

he was found by the early labourers lying on
the ground, near the churchyard, in a state of
insensibility.   All he could tell them was,
that he had followed this lady till she had
turned her head and looked round at him,
when, seized with horror, he had fainted.
"Oh," said they, "you have seen the lady of
Dunn!" What is the legend attached to this
lady of Dunn, I do not know.

A Monsieur De S. had been violently in love
with Hippolyte Clairon, the celebrated French
actress, but she rejected his suit, in so per-
emptory a manner, that even when he was
at the point of death, she refused his earnest
entreaties, that she would visit him. Indignant
at her cruelty, he declared he would haunt
her, and he certainly kept his word. I believe she never saw his ghost, but he appears to have been always near her; at least,
on several occasions when other people doubted
the fact, he signalized his presence at her
bidding, by various sounds, and this, wherever
she happened to be at the moment. Sometimes it was a cry—at others, a shot, and at
others, a clapping of hands or music.   She
seems to have been slow to believe in the
extra-natural character of these noises; and
even when she was ultimately convinced, to have
been divided betwixt horror, on the one hand,

and diversion, at the oddness of the circumstance, on the other. The sounds were heard
by everybody in her vicinity; and I am informed by Mr. Charles Kirkpatrick Sharpe,
that the Margrave of Anspach, who was subsequently her lover, and Mr. Keppel Craven,
were perfectly well acquainted with the circumstances of this haunting, and entertained
no doubt of the facts above alluded to.

The ghost, known by the designation of
" the white lady," which is frequently seen in
different castles or palaces, belonging to the
Royal Family of Prussia, has been mentioned
in another publication, I think. She was
long supposed to be a Countess Agnes of
Orlamunde; but a picture of a princess, called
Bertha, or Perchta von Rosenberg, discovered
some time since, was thought so exceedingly
to resemble the apparition, that it is now a
disputed point which of the two ladies it is;
or whether it is or is not the same apparition
that is seen at different places. Neither of
these ladies appear to have been very happy
in their lives; but the opinion of its being
the Princess Bertha, who lived in the fifteenth
century, was somewhat countenanced by the
circumstance, that at a period when in consequence of the war, an annual benefit which

she had bequeathed to the poor was neglected, the apparition seemed to be unusually disturbed, and was seen more frequently. She is often observed before a death; and one of the Fredericks said, shortly before his decease, that he should "not live long; for he had met the white lady." She wears a widow's band and veil, but it is sufficiently transparent to show her features, which do not express happiness, but placidity, She has only been twice heard to speak. In December, 1628, she appeared in the palace, at Berlin, and was heard to say, "*Veni, judica vivos et mortuos! Judicium mihi adhuc superest*"—Come, judge the quick and the dead! I wait for judgment.

On the other occasion, which is more recent, one of the princesses at the Castle of Neuhaus, in Bohemia, was standing before a mirror, trying on a new head-dress, when on asking her waiting-maid, what the hour was, the white lady suddenly stept from behind a screen and said, "Zehn uhr ist es ihr Liebden! —It is ten o'clock, your love!" which is the mode in which the sovereign princes address each other, instead of "your highness." The princess was much alarmed; soon fell sick, and died in a few weeks. She has frequently

evinced displeasure at the exhibition of im-
piety or vice ; and there are numerous records
of her different appearances to be found in
the works of Balbinus, and of Erasmus
Francisci ; and in a publication called " The
Iris," published in Frankfort, in 1819, the
editor, George Doring, who is said to have
been a man of great integrity, gives the following account of one of her later appearances,
which he declares he received just as he gives
it, from the lips of his own mother, on whose
word and judgment he could perfectly rely ;
and shortly before his death, an enquiry being
addressed to him with regard to the correct-
ness of the narration, he vouched for its
authenticity.

It seems that the elder sister of his mother
was companion to one of the ladies of the
court, and that the younger ones were in the
habit of visiting her frequently. Two of these
(Doring's mother and another) aged fourteen
and fifteen, were once spending a week with
her, when she being out and they alone with
their needle-work, chattering about the court
diversions, they suddenly heard the sound of a
stringed instrument, like a harp, which seemed
to proceed from behind a large stove, that
occupied one corner of the room.   Half in fear

and half in fun, one of the girls took a yard measure that lay beside them, and struck the spot, whereupon, the music ceased, but the stick was wrested from her hand. She became alarmed; but the other, named Christina, laughed, and said, she must have fancied it, adding, that the music, doubtless, proceeded from the street, though they could not descry any musicians. To get over her fright, of which she was half ashamed, the former now ran out of the room, to visit a neighbour for a few minutes, but when she returned, she found Christina lying on the floor, in a swoon; who, on being revived, with the aid of the attendants who had heard a scream, related, that no sooner had her sister left her, than the sound was repeated, close to the stove, and a white figure had appeared, and advanced towards her, whereupon, she had screamed and fainted.

The lady who owned the apartments, flattered herself that this apparition betokened that a treasure was hidden under the stove, and, imposing silence on the girls, she sent for a carpenter, and had the planks lifted. The floor was found to be double, and below was a vault, from which issued a very unwholesome vapour, but no treasure was found, nor anything but a quantity of quick lime. The

circumstance being now made known to the
King, he expressed no surprise; he said that
the apparition was doubtless that of a Countess
of Orlamunde, who had been built up alive in
that vault. She was the mistress of a Margrave of Brandenburg, by whom she had two
sons. When the prince became a widower,
she expected he would marry her; but he
urged, as an objection, that he feared, in that
case, her sons might hereafter dispute the
succession with the lawful heirs. In order to
remove this obstacle out of her way, she
poisoned the children; and the Margrave,
disgusted and alarmed, had her walled up in
that vault for her pains. He added that she
was usually seen every seven years, and was
preceded by the sound of a harp, on which
instrument she had been a proficient; and
also that she more frequently appeared to
children than to adults, as if the love she had
denied her own offspring in life was now her
torment, and that she sought a reconciliation
with childhood in general. I know from the
best authority that the fact of these appearances is not doubted by those who have the
fullest opportunities of enquiry and investigation; and I remember seeing in the English
papers, a few years since, a paragraph copied

from the foreign journals, to the effect that the
White Lady had been seen again, I think at
Berlin.

The following very curious relation, I have
received from the gentleman to whom the circumstance occurred, who is a professional man,
residing in London.

"I was brought up by a grandfather and four
aunts, all ghost-seers, and believers in supernatural appearances. The former had been a
sailor, and was one of the crew that sailed
round the world with Lord Anson. I remember when I was about eight years old, that
I was awakened by the screams of one of these
ladies, with whom I was sleeping, which summoned all the family about her, to enquire the
cause of the disturbance. She said, that she
had 'Seen Nancy by the side of the bed, and
that she was slipping into it.' We had scarcely
got down stairs in the morning, before intelligence arrived, that that lady had died, precisely at the moment my aunt said she saw
her. Nancy was her bother's wife. Another
of my aunts, who was married and had a
large family, foretold my grandfather's death,
at a time that we had no reason to apprehend
it ; he also had appeared at her bed side. He
was then alive and well ; but he died a fortnight

afterwards. But it would be tedious were I to enumerate half the instances I could recall of a similar description ; and I will therefore proceed to the relation of what happened to myself.

" I was, some few years since, invited to pass a day and night at the house of a friend in Hertfordshire, with whom I was intimately acquainted. His name was B., and he had formerly been in business as a saddler, in Oxford-street, where he had realised a handsome fortune, and had now retired to enjoy his *otium cum dignitate*, in the rural and beautiful village of Sarratt.

" It was a gloomy Sunday, in the month of November, when I mounted my horse for the journey, and there was so much appearance of rain, that I should certainly have selected some other mode of conveyance, had I not been desirous of leaving the animal in Mr. B.'s straw-yard for the winter. Before I got as far as St. John's Wood, the threatening clouds broke, and, by the time I reached Watford, I was completely soaked. However, I proceeded, and arrived at Sarratt before my friend and his wife had returned from church. The moment they did so, they furnished me with dry clothes, and I was informed that we were to dine at the house of Mr. D., a very agreeable

neighbour.   I felt some little hesitation about
presenting myself in such a costume, for I was
decked out in a full suit of Mr. B.'s, who was
a stout man, of six feet in height, whilst I am
rather of the diminutive order; but my objections were over-ruled; we went, and my appearance added not a little to the hilarity of
the party.   At ten o'clock we separated, and I
returned with Mr. and Mrs. B. to their house,
where I was shortly afterwards conducted to
a very comfortable bed-room.

"Fatigued with my day's ride, I was soon in
bed, and soon asleep, but I do not think I could
have slept long, before I was awakened by the
violent barking of dogs.   I found that the
noise had disturbed others as well as myself,
for I heard Mr. B., who was lodged in the
adjoining room, open his window and call to
them to be quiet.   They were obedient to his
voice, and as soon as quietness ensued, I dropt
asleep again; but I was again awakened by
an extraordinary pressure upon my feet; *that
I was perfectly awake, I declare;* the light
that stood in the chimney-corner shone
strongly across the foot of the bed, and I saw
the figure of a well-dressed man in the act of
stooping, and supporting himself in so doing
by the bed-clothes.   He had on a blue coat,

with bright gilt buttons, but I saw no head;
the curtains at the foot of the bed, which were
partly looped back, just hung so as to conceal
that part of his person. At first, I thought it
was my host, and as I had dropt my clothes,
as is my habit, on the floor, at the foot of the
bed, I supposed he was come to look after them,
which rather surprised me: but, just as I had
raised myself upright in bed, and was about
to enquire into the occasion of his visit, the
figure passed on. I then recollected that I
had locked the door; and, becoming somewhat puzzled, I jumped out of bed; but
I could see nobody; and on examining
the room, I found no means of ingress
but the door through which I had entered,
and one other; both of which were locked on
the inside. Amazed and puzzled I got into
bed again, and sat some time ruminating on
the extraordinary circumstance, when it
occurred to me that I had not looked under
the bed. So I got out again, fully expecting
to find my visitor, whoever he was, there; but
I was disappointed. So after looking at my
watch, and ascertaing that it was ten minutes
past two, I stept into bed again, hoping now
to get some rest. But, alas! sleep was
banished for that night; and after turning

from side to side, and making vain endeavours
at forgetfulness, I gave up the point, and lay
till the clocks struck seven, perplexing my
brain with the question of who my midnight
visitor could be; and also how he had got in
and how he had got out of my room. About
eight o'clock, I met my host and his wife at
the breakfast-table, when, in answer to their
hospitable enquires of how I had passed the
night, I mentioned, first, that I had been
awaked by the barking of some dogs, and that
I had heard Mr. B. open his window and call
to them. He answered that two strange dogs
had got into the yard and had disturbed the
others. I then mentioned my midnight
visitor, expecting that they would either explain the circumstance, or else laugh at me
and declare I must have dreamt it. But, to
my surprise, my story was listened to with
grave attention; and they related to me the
tradition with which this spectre, for such I
found they deemed it to be, was supposed to
be connected. This was to the effect, that
many years ago, a gentleman, so attired, had
been murdered there, under some frightful
circumstances; and that his head had been
cut off. On perceiving that I was very unwilling to accept this explanation of the

mystery, for in spite of my family peculiarity,
I had always been an entire disbeliever in
supernatural appearances, they begged me to
prolong my visit for a day or two, when they
would introduce me to the rector of the parish,
who could furnish me with such evidence with
regard to circumstances of a similar nature, as
would leave no doubt on my mind as to the
possibility of their occurrence. But I had made
an engagement to dine at Watford, on my way
back ; and I confess, moreover, that after what
I had heard, I did not feel disposed to encounter the chance of another visit from the
mysterious stranger; so I declined the proffered
hospitality and took my leave.

" Some time after this, I happened to be
dining in C--- Street, in company with some
ladies resident in the same county, when
chancing to allude to my visit to Sarratt, I
added, that I had met with a very extraordinary adventure there, which I had never been
able to account for; when one of these ladies
immediately said, that she hoped I had not
had a visit from the headless gentleman, in a
blue coat and gilt buttons, who was said to
have been seen by many people in that house.

" Such is the conclusion of this marvellous
tale as regards myself; and I can only assure

you that I have related facts as they occurred;
and that I had never heard a word about this
apparition in my life, till Mr. B., related to me
the tradition above alluded to. Still, as I am
no believer in supernatural appearances, I am
constrained to suppose that the whole affair
was the product of my imagination.

"I must add, that Mr. B. mentioned some
strange circumstances connected with another
house in the county, inhabited by a Mr. M.,
which were corroborated by the ladies above
alluded to. Both parties agreed that, from
the unaccountable noises, &c. &c., which were
heard there, that gentleman had the greatest
difficulty in persuading any servants to remain
with him.

"(Signed)   A. W. M.
"C— Street,
"5th September, 1846."

This is one of those curious . instances of
determined scepticism that fully justify the
patriarch's prediction.

The following interesting letter, written by
a member of a very distinguished English
family, will furnish its own explanation :—

"As you express a wish to know what degree of credit is to be attached to a garbled

tale, which has been sent forth, after a lapse
of between thirty and forty years, as an 'accredited ghost-story,' I will state the facts
as they were recalled to my mind last year, by
a daughter of Sir William A. C., who sent the
book to me, requesting me to tell her if there
was any foundation for the story, which she
could scarcely believe, since she had never
heard my mother allude to it. I read the narrative with surprise, it being evidently not furnished by any of the family, nor indeed by
any one who was with us at the time! yet
though full of mistakes in names, &c. &c.
some particulars come so near the truth as to
puzzle me. The facts are as follows :—

"Sir James, my mother, with myself and my
brother Charles, went abroad towards the end
of the year 1786. After trying several different places, we determined to settle at Lille,
where we found the masters particularly good,
and where we had also letters of introduction
to several of the best French families. There
Sir James left us, and, after passing a few
days in an uncomfortable lodging, we engaged
a nice large family house, which we liked
much, and which we obtained at a very low
rent, even for that part of the world.

"About three weeks after we were esta-

blished in our new residence, I walked one
day, with my mother to the bankers, for the
purpose of delivering our letter of credit from
Sir Robert Herries, and drawing some money,
which being paid in heavy five-franc pieces,
we found we could not carry, and therefore requested the banker to send, saying, 'We live
in the Place Du Lion D'or.' Whereupon, he
looked surprised, and observed that he knew
of no house there fit for us, 'Except, indeed,'
he added, 'the one that has been long uninhabited, on account of the *revenant* that walks
about it. He said this quite seriously, and in
a natural tone of voice; in spite of which we
laughed, and were quite entertained at the idea
of a ghost; but, at the same time, we begged
him not to mention the thing to our servants,
lest they should take any fancies into their
heads; and my mother and I resolved to say
nothing about the matter to any one. 'I suppose it is the ghost,' said my mother, laughing,
' that wakes us so often by walking over our
heads.' We had, in fact, been awakened several nights, by a heavy foot, which we supposed to be that of one of the men servants,
of whom we had three English and four
French; of women servants we had five
English, and all the rest were French. The

English ones, men and women, every one of them, returned ultimately to England with us.

"A night or two afterwards, being again awakened by the step, my mother asked Creswell, 'Who slept in the room above us?' 'No one, my lady,' she replied, 'It is a large empty garret.'

"About a week or ten days after this, Creswell came to my mother, one morning, and told her that all the French servants talked of going away, because there was a *revenant* in the house; adding, that there seemed to be a strange story attached to the place, which was said, together with some other property, to have belonged to a young man, whose guardian, who was also his uncle, had treated him cruelly, and confined him in an iron cage; and as he had subsequently disappeared, it was conjectured he had been murdered. This uncle, after inheriting the property, had suddenly quitted the house, and sold it to the father of the man of whom we had hired it. Since that period, though it had been several times let, nobody had ever staid in it above a week or two; and, for a considerable time past it had had no tenant at all.

"'And do you really believe all this nonsense, Creswell?' said my mother.

" ' Well, I don't know, my lady,' answered she; ' but there's the iron cage in the garret over your bed-room, where you may see it, if you please.'

" Of course we rose to go, and as just at that moment an old officer, with his Croix de St. Louis, called on us; we invited him to accompany us, and we ascended together. We found, as Creswell had said, a large empty garret, with bare brick walls, and in the further corner of it stood an iron cage, such as wild beasts are. kept in, only higher; it was about four feet square, and eight in height, and there was an iron ring in the wall at the back, to which was attached an old rusty chain, with a collar fixed to the end of it. I confess it made my blood creep, when I thought of the possibility of any human being having inhabited it! And our old friend expressed as much horror as ourselves, assuring us that it must certainly have been constructed for some such dreadful purpose. As, however, we were no believers in ghosts, we all agreed that the noises must proceed from somebody who had an interest in keeping the house empty; and since it was very disagreeable to imagine that there were secret means of entering it by night, we resolved, as soon as possible, to look out for another residence, and, in the mean time, to say nothing about the matter to anybody. About ten days after this determination, my mother, observing one morning that Creswell, when she came to dress her, looked exceedingly pale and ill, enquired if anything was the matter with her ? 'Indeed, my lady,' answered she, 'we have been frightened to death ; and neither I nor Mrs. Marsh can sleep again in the room we are now in.'

" ' Well,' returned my mother, 'you shall both come and sleep in the little spare room next us; but what has alarmed you?'

" ' Some one, my lady, went through our room in the night ; we both saw the figure, but we covered our heads with the bed-clothes, and lay in a dreadful fright till morning.'

" On hearing this, I could not help laughing, upon which Creswell burst into tears ; and seeing how nervous she was, we comforted her, by saying, we had heard of a good house, and that we should very soon abandon our present habitation.

" A few nights afterwards, my mother requested me and Charles to go to her bed-room, and fetch her frame, that she might prepare her work for the next day. : It was after sup-

per; and we were ascending the stairs by the
light of a lamp which was always kept burning, when we saw going up before us, a tall,
thin figure, with hair flowing down his back,
and wearing a loose powdering gown. We
both at once concluded it was my sister
Hannah, and called out, 'It won't do, Hannah!
You cannot frighten us!' Upon which the
figure turned into a recess in the wall; but as
there was nobody there, when we passed, we
concluded that Hannah had contrived, somehow or other, to slip away and make her
escape by the back stairs. On telling this to
my mother, however, she said, 'It is very odd!
for Hannah went to bed with a head-ache before you came in from your walk;' and sure
enough, on going to her room, there we found
her fast asleep; and Alice, who was at work
there, assured us that she had been so for
more than an hour. On mentioning this circumstance to Creswell, she turned quite pale,
and exclaimed that that was precisely the
figure she and Marsh had seen in their bed-
room.

"About this time, my brother Harry came
to spend a few days with us, and we gave him
a room up another pair of stairs, at the opposite end of the house. A morning or two after

his arrival, when he came down to breakfast, he asked my mother, angrily, whether she thought he went to bed drunk and could not put out his own candle, that she sent those French rascals to watch him. My mother assured him that she had never thought of doing such a thing; but he persisted in the accusation, adding, 'Last night I jumped up and opened the door, and by the light of the moon, through the skylight, I saw the fellow in his loose gown at the bottom of the stairs. If I had not been in my shirt, I would have gone after him and made him remember coming to watch me.'

"We were now preparing to quit the house, having secured another, belonging to a gentleman who was going to spend some time in Italy; but a few days before our removal, it happened, that a Mr. and Mrs. Atkyns, some English friends of ours, called, to whom we mentioned these strange circumstances, observing, how extremely unpleasant it was, to live in a house that somebody found means of getting into, though how they contrived it we could not discover, nor what their motive could be, except it was to frighten us; observing, that nobody could sleep in the room Marsh and Creswell had been obliged to give

up. Upon this, Mrs. Atkyns laughed heartily,
and said, that she should like, of all things,
to sleep there, if my mother would allow her,
adding, that, with her little terrier, she should
not be afraid of any ghost that ever appeared.
As my mother had, of course, no objection to
this fancy of hers, she requested Mr. Atkyns
to ride home with the groom, in order that
the latter might bring her night-things before
the gates of the town were shut, as they were
then residing a little way in the country. Mr.
Atkyns smiled, and said she was very bold;
but he made no difficulties, and sent the things,
and his wife retired with her dog to her room,
when we retired to ours, apparently without
the least apprehension.

"When she came down in the morning
we were immediately struck at seeing her look
very ill; and, on enquiring if she, too, had
been frightened, she said she had been
awakened in the night by something moving
in her room, and that, by the light of the
night-lamp, she saw, most distinctly, a figure,
and that the dog, which was very spirited
and flew at everything, never stirred, although
she had endeavoured to make him. We saw
clearly that she had been very much alarmed ;
and when Mr. Atkyns came, and endeavoured

to dissipate the feeling by persuading her that she might have dreamt it, she got quite angry. We could not help thinking that she had actually seen something; and my mother said, after she was gone, that, though she could not bring herself to believe it was really a ghost, still she earnestly hoped that she might get out of the house without seeing this figure, which frightened people so much.

"We were now within three days of the one fixed for our removal; I had been taking a long ride, and, being tired, had fallen asleep the moment I lay down, but, in the middle of the night, I was suddenly awakened—I cannot tell by what, for the step over our heads we had become so used to that it no longer disturbed us. Well, I awoke; I had been lying with my face towards my mother, who was asleep beside me, and, as one usually does on awaking, I turned to the other side, where, the weather being warm, the curtain of the bed was undrawn, as it was, also, at the foot, and I saw, standing by a chest of drawers, which were betwixt me and the window, a thin, tall figure, in a loose powdering gown, one arm resting on the drawers, and the face turned towards me. I saw it quite distinctly by the night-light, which burnt clearly; it was a long, thin, pale, young face, with, oh, such a melancholy expression, as can never be effaced from my memory ! I was, certainly, very much frightened ; but my great horror was, lest my mother should awake and see the figure. I turned my head gently towards her, and heard her breathing high in a sound sleep. Just then the clock on the stairs struck four. I dare say it was nearly an hour before I ventured to look again, and when I did take courage to turn my eyes towards the drawers, there was nothing, yet I had not heard the slightest sound, though I had been listening with the greatest intensity.

"As you may suppose, I never closed my eyes again; and glad I was when Creswell knocked at the door, as she did every morning, for we always locked it, and it was my business to get out of bed and let her in. But, on this occasion, instead of doing so, I called out, ' Come in; the door is not fastened;' upon which she answered that it was, and I was obliged to get out of bed and admit her as usual.

"When I told my mother what had happened, she was very grateful to me for not waking her, and commended me much for my resolution; but as she was always my first object, that was not to be wondered at. She however resolved not to risk another night in the house; and we got out of it that very day, after instituting, with the aid of the servants, a thorough search, with a view to ascertain if there was any possible means of getting into the rooms except by the usual modes of ingress; but our search was vain; none could be discovered.

" I think, from the errors in the names, &c. that the publisher of the " Accredited Ghost Stories" must have obtained his account from the inhabitants of Lille."

Considering the number of people that were in the house, the fearlessness of the family, and their disinclination to believe in what is called *the supernatural,* together with the great interest the owner of this large and handsome residence must have had in discovering the trick, if there had been one, I think it is difficult to find any other explanation of this strange story, than that the sad and disappointed spirit of this poor injured, and probably murdered boy, had never been disengaged from its earthly relations, to which regret for its frustrated hopes and violated rights, still held it attached.

There is a story told by Pliny, the

younger, of a house at Athens, in which nobody could live, from its being haunted.   At
length, the philosopher Athenadorus took it;
and the first night he was there, he seems to
have comported himself very much, as the
courageous Mrs. Canning did on a similar
occasion, at Plymouth. He sent his servants
to bed, and set himself seriously to work with
his writing materials, determined that fancy
should not be left free to play him false. For
some time all was still, and his mind was
wholly engaged in his labours, when he heard
a sound like the rattling of chains—which
was the sound that had frightened everybody
out of the house; but Athenadorus closed his
ears, kept his thoughts collected, and wrote on,
without lifting up his eyes. The noise, however increased; it approached the door; it
entered the room; then he looked round, and
beheld the figure of an old man, lean, haggard,
and dirty, with dishevelled hair, and a long
beard, who held up his finger and beckoned
him. Athenadorus made a gesture with his
own hand, in return, signifying that he should
wait, and went on with his writing. Then,
the figure advanced and shook his chains over
the philosopher's head, who, on looking up,
saw him beckoning as before; whereupon he
arose and followed him. The apparition

walked slowly, as if obstructed by his chains, and having conducted him to a certain spot in the court, which separated the two divisions of an ancient Greek house, he suddenly disappeared. Athenadorus gathered together some grass and leaves, in order to mark the place, and the next day he recommended the authorities to dig there; which they did, and found the skeleton of a human being encircled with chains. It being taken up, and the rights of sepulture duly performed, the house was no longer disturbed.

This was, probably, some poor prisoner also; and in his desire to direct notice to his body, we see the prejudices of his age and country surviving dissolution. Grose the antiquary, who is, as I have before observed, very facetious on the subject of ghosts, remarks that "Dragging chains is not the custom of English ghosts, chains and black vestments being chiefly the accoutrements of foreign spectres, seen in arbitrary governments." Now, this is a very striking observation. Grose's studies, had, doubtless, introduced him to many histories of this description; and the different characteristics of these apparitions under different governments, is a circumstance in remarkable conformity with the views of

those who have been led to take a much more
serious view of the subject. They appear as
they lived, and as they conceive of themselves;
and when rapport or receptivity enable them
to see, and to render themselves visible to those
yet living in the flesh, it is by so appearing
that they tell their story, and ask for sympathy
and assistance. I say enable them *to see*, because there seem many reasons for concluding
that they do not, under ordinary circumstances, see us, any more than we see them.
Whether it be rapport with certain inhabitants,
or whether the phenomenon be dependent on
certain periods, or any other condition, we
cannot tell; but I have met with several
accounts of houses in which an annoyance of
this sort has recurred more than once, at
different intervals, sometimes at a distance of
seven or ten years, the intermediate time being
quite free from it.

One of the most melancholy and impressive
circumstances of this sort I have met with,
occurred to Mrs. L., a lady with whose family
I am acquainted; Mrs. L. herself having been
kind enough to furnish me with the particulars :—A few years since, she took a furnished
house in Stevenson Street, North Shields, and
she had been in it a very few hours, before she

was perplexed by hearing feet in the passage,
though, whenever she opened the door, she
could see nobody. She went to the kitchen,
and asked the servant if she had not heard
the same sound; she said she had not, but
that there seemed to be strange noises in the
house. When Mrs. L. went to bed, she could
not go to sleep for the noise of a child's rattle,
which seemed to be inside her curtains. It
rattled round her head, first on one side then
on the other; then there were sounds of feet
and of a child crying, and a woman sobbing;
and, in short, so many strange noises, that
the servant became frightened, and went
away. The next girl Mrs. L. engaged came
from Leith, and was a stranger to the place;
but she had only passed a night in the house,
when she said to her mistress, "This is a
troubled house you've got into, Ma'am," and
she described, amongst the rest, that she had
repeatedly heard her own name called by a
voice near her, though she could see nobody.

One night Mrs. L. heard a voice, like nothing
human, close to her, cry, "Weep! Weep!
Weep!" Then there was a sound like some
one struggling for breath, and again, "Weep!
Weep! Weep!" Then the gasping, and a
third time, "Weep! Weep! Weep!" She

stood still, and looked steadfastly on the spot whence the voice proceeded, but could see nothing; and her little boy, who held her hand, kept saying, " What is that, Mamma ? What is that ?" She describes the sound as most frightful. All the noises seemed to suggest the idea of childhood, and of a woman in trouble. One night, when it was crying round her bed, Mrs. L. took courage and adjured it; upon which the noise ceased, for that time, but there was no answer. Mr. L. was at sea when she took the house, and when he came home, he laughed at the story at first, but soon became so convinced the account she gave was correct, that he wanted to have the boards taken up, because from the noises seeming to hover much about one spot, he thought perhaps some explanation of the mystery might be found. But Mrs. L. objected that if anything of a painful nature were discovered she should not be able to continue in the house; and as she must pay the year's rent, she wished, if possible, to make out the time.

She never saw anything but twice; once, the appearance of a child seemed to fall from the ceiling, close to her, and then disappear; and another time she saw a child run into a closet in a room at the top of the house ; and it was most remarkable that a small door in that room, which was used for going out out the roof, always stood open.   However often they shut it, it was opened again immediately by an unseen hand, even before they got out of the room, and this continued the whole time they were in the house ; whilst night and day, some one in creaking shoes was heard pacing backwards and forwards in the room over Mr. and Mrs. L.'s head.

At length the year expired ; and to their great relief they quitted the house : but five or six years afterwards, a person who had bought it having taken up the floor of that upper room to repair it, there was found, close to the small door above alluded to, the skeleton of a child.   It was then remembered, that some years before, a gentleman of somewhat dissolute habits, had resided there ; and that he was supposed to have been on very intimate terms with a young woman servant, who lived with him ; but there had been no suspicion of anything more criminal.

About six years ago, Mr. C., a gentleman, engaged in business in London, heard of a good country house in the neighbourhood of the metropolis, which was to be had at a low

rent. It was rather an old-fashioned place,
and was surrounded by a garden and pleasure-
ground; and having taken a lease of it for
seven years, furnished as it was, his family removed thither, and he joined them once or
twice a week, as his business permitted.

They had been some considerable time in
the house without the occurrence of anything
remarkable, when one evening, towards dusk,
Mrs. C., on going into what was called the oak
bed-room, saw a female figure near one of the
windows. It was apparently a young woman
with dark hair hanging over her shoulders, a
silk petticoat, and a short white robe, and she
appeared to be looking eagerly through the
window, as if expecting somebody. Mrs. C.
clapped her hand upon her eyes " as thinking
she had seen something she ought not to have
seen," and when she looked again, the figure
had disappeared.

Shortly after this, a young girl who filled
the situation of under nursery-maid, came to
her in great agitation, saying, that she had
had a terrible fright, from seeing a very ugly
old woman looking in upon her as she passed
the window in the lobby. The girl was trembling violently, and almost crying, so that Mrs.
C. entertained no doubts of the reality of her

alarm. She, however, thought it advisable to
laugh her out of her fear, and went with her
to the window, which looked into a closed
court, but there was no one there; neither had
any of the other servants seen such a person.
Soon after this, the family began to find
themselves disturbed with strange and frequently very loud noises, during the night.
Amongst the rest, there was something like
the beating of a crow-bar upon the pump in
the above-mentioned court; but, search as
they would, they could discover no cause for
the sound. One day, when Mr. C. had brought
a friend from London to stay the night with
him, Mrs. C. thought proper to go up to the
oak bed-room, where the stranger was to sleep,
for the purpose of inspecting the arrangements
for his comfort, when, to her great surprise,
some one seemed to follow her up to the fire
place, though, on turning round, there was nobody to be seen. She said nothing about it, however, and returned below, where her husband,
and the stranger were sitting.   Presently, one
of the servants (not the one mentioned above)
tapped at the door, and requested to speak with
her, and Mrs. C. going out, she told her, in
great agitation, that in going up stairs to the
visitor's room, a footstep had followed all the

way to the fire-place, although she could see
nobody.   Mrs. C. said something soothing,
and that matter passed, she, herself, being a
good deal puzzled, but still unwilling to admit
the idea that there was anything extra-natural
in these occurrences.   Repeatedly, after this,
these foot-steps were heard in different parts
of the house, when nobody was to be seen;
and often, whilst she was lying in bed, she
heard them distinctly approach her door, when,
being a very courageous woman, she would
start out with a loaded pistol in her hand, but
there was never any one to be seen.   At length
it was impossible to conceal from herself and
her servants that these occurrences were of an
extraordinary nature, and the latter, as may
be supposed, felt very uncomfortable.   Amongst
other unpleasant things, whilst sitting all
together in the kitchen, they used to see the
latch lifted and the door open, though no one
came in that they could see; and when Mr.
C. himself watched for these events, although
they took place, and he was quite on the alert,
he altogether failed in detecting any visible
agent.

One night, the same servant who had heard
the footsteps following her to the bed-room fire-
place, happening to be asleep in Mrs. C.'s

chamber, she became much disturbed, and
was heard to murmur " Wake me ! Wake me !"
as if in great mental anguish. Being aroused,
she told her mistress a dream she had had,
which seemed to throw some light upon these
mysteries. She thought she was in the oak
bed-room, and at one end of it she saw a young
female in an old fashioned dress, with long
dark hair; whilst in another part of the room,
was a very ugly old woman, also in old-
fashioned attire. The latter addressing the
former, said, " What have you done with the
child, Emily ? What have you done with the
child ? " To which the younger figure
answered, " Oh, I did not kill it. He was preserved, and grew up, and joined the —— Regiment, and went to India." Then addressing
the sleeper, the young lady continued, " I have
never spoken to mortal before; but I will tell
you all. My name is Miss Black; and this
old woman is Nurse Black. Black is not her
name ; but we call her so, because she has
been so long in the family." Here the old
woman interrupted the speaker by coming up
and laying her hand on the dreaming girl's
shoulder, whilst she said something; but she
could not remember what, for feeling excruci-
ating pain from the touch, she had been so far

aroused as to be sensible she was asleep, and to beg to be wholly awakened.

As the old woman seemed to resemble the figure that one of the other servants had seen looking into the window, and the young one resembled that she had herself seen in the oak chamber, Mrs. C. naturally concluded that there was something extraordinary about this dream; and she consequently took an early opportunity of enquiring in the neighbourhood what was known as to the names or circumstances of the former inhabitants of this house; and after much investigation she learnt, that about seventy or eighty years before, it had been in the possession of a Mrs. Ravenhall, who had a niece, named Miss Black, living with her. This niece Mrs. C. supposed might be the younger of the two persons who had been seen. Subsequently, she saw her again in the same room, wringing her hands, and looking with a mournful significance to one corner. They had the boards taken up on that spot; but nothing was found.

One of the most curious incidents connected with this story, remains to be told. After occupying the house three years, they were preparing to quit it—not on account of its being haunted, but for other reasons—when, on

awaking one morning, a short time before
their departure, Mrs. C. saw standing at the
foot of her bed, a dark complexioned man, in
a working dress, a fustian jacket, and red com-
forter round his neck, who, however, suddenly
disappeared. Mr. C. was lying beside her at the
time, but asleep. This was the last apparition
that was seen; but the strange thing is, that a
few days after this, it being necessary to order
in a small quantity of coals, to serve till their
removal, Mr. C. undertook to perform the
commission on his way to London. Accordingly, the next day she mentioned to him, that
the coals had arrived; which he said was very
fortunate, since he had entirely forgotten to
order them. Wondering whence they had
come, Mrs. C. hereupon, enquired of the servants, who none of them knew anything about
the matter; but, on interrogating a person
in the village, with whom they had frequently
been provided with this article, he answered,
that they had been ordered by a dark man, in
a fustian jacket and red comfort, who had
called for the purpose !

After this last event, Mr. and Mrs. C. quitted
the house; but I have heard that its subsequent tenants encountered some similar annoyances, although I have no means of ascertaining
the particulars.

But, perhaps, one of the most remarkable cases of haunting in modern times, is that of Willington, near Newcastle, in my account of which, however, I find myself anticipated by Mr. Howitt; and as he has had the advantage of visiting the place, which I have not, I shall take the liberty of borrowing his description of it, prefacing the account with the following letter from Mr. Procter, the owner of the house, who, it will be seen, vouches for the general authenticity of the narrative. The letter was written in answer to one from me, requesting some more precise information than I had been able to obtain.

" Josh. Proctor, hopes C. Crowe will excuse her note having remained two weeks unanswered, during which time, J. P. has been from home, or particularly engaged. Feeling averse to add to the publicity the circumstances occurring in his house, at Willington, have already obtained, J. P. would rather not furnish additional particulars; but if C. C. is not in possession of the number of ' Howitt's Journal,' which contains a variety of details on the subject, he will be glad to forward her one. He would at the same time, assure C. Crowe of the strict accuracy, of that portion of W. Howitt's narrative, which is extracted from

' Richardson's Table Book.' W. Howitt's state-
ments derived from his recollection of verbal
communications, with branches of J. Procter's
family, are likewise essentially correct, though,
as might be expected in some degree, erro-
neous circumstantially.

" J. P. takes leave to express his conviction,
that the unbelief of the educated classes, in
apparitions of the deceased and kindred phenomena, is not grounded on a fair philosophic
examination of the facts, which have induced
the popular belief of all ages and countries ;
and that it will be found by succeeding ages, to
have been nothing better than unreasoning and
unreasonable prejudice.

" Willington, near Newcastle-on-Tyne,
7th mo. 22, 1847."

" VISITS TO REMARKABLE PLACES.

" BY WILLIAM HOWITT.

" THE HAUNTED HOUSE AT WILLINGTON, NEAR
NEWCASTLE-ON-TYNE.

" We have of late years settled it as an
established fact, that ghosts and haunted
houses were the empty creation of ignorant
times. We have comfortably persuaded ourselves that such fancies only hovered in the

twilight of superstition, and that in these enlightened days they had vanished for ever.
How often has it been triumphantly referred
to, as a proof that all such things were the
offspring of ignorance—that nothing of the
kind is heard of now? What shall we say,
then, to the following facts? Here we have
ghosts, and a haunted house still. We have
them in the face of our vaunted noon-day
light, in the midst of a busy and a populous
neighbourhood, in the neighbourhood of a
large and most intelligent town, and in a
family neither ignorant, nor in any other
respect superstitious. For years have these
ghosts and hauntings disturbed the quiet
of a highly respectable family, and continue
to haunt and disturb, spite of the incredulity
of the wise, the investigations of the curious,
and the anxious vigilance of the suffering
family itself.

" Between the railway running from Newcastle-on-Tyne to North Shields, and the
river Tyne, there lies in a hollow some few
cottages, a parsonage, and a mill and a miller's
house. These constitute the hamlet of Willington. Just above these the railway is carried
across the valley on lofty arches, and from it
you look down on the mill and cottages, lying

at a considerable depth below. The mill is a large steam flour mill, like a factory, and the miller's house stands near it, but not adjoining it. None of the cottages which lie between these premises and the railway, either, are in contact with them. The house stands on a sort of little promontory, round which runs the channel of a water-course, which appears to fill and empty with the tides. On one side of the mill and house, slopes away, upwards, a field to a considerable distance, where it is terminated by other enclosures; on the other stands a considerable extent of ballast-hill, *i. e.*, one of the numerous hills on the banks of the Tyne, made by the deposit of ballast from the vessels trading thither. At a distance, the top of the mill seems about level with the country around it. The place lies about half-way bewteen Newcastle and North Shields.

" This mill is, I believe, the property of, and is worked by, Messrs. Unthank and Procter. Mr. Joseph Procter resides on the spot in the house just by the mill, as already stated. He is a member of the Society of Friends, a gentleman in the very prime of life; and his wife, an intelligent lady, is of a family of Friends in Carlisle. They have several young children.

This very respectable and well-informed family,
belonging to a sect which of all others is most
accustomed to controul, to regulate, and to put
down even the imagination; the last people in
the world, as it would appear, in fact, to be
affected by any mere imaginary terrors or impressions, have for years been persecuted by
the most extraordinary noises and apparitions.

"The house is not an old house, as will
appear; it was built about the year 1800. It
has no particularly spectral look about it. Seeing it in passing, or within, ignorant of its real
character, one should by no means say that it
was a place likely to have the reputation of
being haunted. Yet looking down from the
railway, and seeing it and the mill lying in a
deep hole, one might imagine various strange
noises likely to be heard in such a place in the
night, from vessels on the river, from winds
sweeping and howling down the gulley in
which it stands, from engines in the neighbourhood connected with coal mines, one of which,
I could not tell where, was making, at the
time I was there, a wild sighing noise, as I
stood on the hill above. There is not any
passage, however, known of under the house,
by which subterraneous noises could be heard,

nor are they merely noises that are heard ; distinct apparitions are declared to be seen.

"Spite of the unwillingness of Mr. Procter that these mysterious circumstances should become quite public, and averse as he is to make known himself these strange visitations, they were of such a nature that they soon became rumoured over the whole neighbourhood. Numbers of people hurried to the place to enquire into the truth of them, and at length a remarkable occurrence brought them into print. What this occurrence was, the pamphlet which appeared, and which was afterwards reprinted in 'The Local Historian's Table-Book,' published by Mr. M. A. Richardson, of Newcastle, and which I here copy, will ex plain. It will be seen that the writer of this article has the fullest faith in the reality of what he relates, as, indeed, vast numbers of the best informed inhabitants of the neighbourhood have.

"AUTHENTIC ACCOUNT OF A VISIT TO THE HAUNTED HOUSE AT WILLINGTON.

"Were we to draw an inference from the number of cases of reported visitations from the invisible world that have been made public of late, we might be led to imagine that the

days of supernatural agency were about to re-
commence, and that ghosts and hobgoblins
were about to resume their sway over the fears
of mankind. Did we, however, indulge such
an apprehension, a glance at the current tone
of the literature and philosophy of the day,
when treating of these subjects, would show a
measure of unbelief regarding them as scorn-
ful and uncompromising as the veriest atheist
or materialist could desire. Notwithstanding
the prevalence of this feeling amongst the educated classes, there is a curiosity and interest
manifested in every occurrence of this nature,
that indicates a lurking faith at bottom, which
an affected scepticism fails entirely to conceal.
We feel, therefore, that we need not apologise
to our readers for introducing the following
particulars of a *visit* to a house in this immediate neighbourhood, which had become notorious for some years previous, as being
'haunted;' and several of the reputed deeds,
or misdeeds, of its supernatural visitant had
been published far and wide by rumour's
thousand tongues. We deem it as worthy to
be chronicled as the doings of its contem-
porary *genii* at Windsor, Dublin, Liverpool,
Carlisle, and Sunderland, and which have all
likewise hitherto failed, after public investi-

gation, to receive a solution consistent with a rejection of spiritual agency.

" We have visited the house in question, which is well known to many of our readers as being near a large steam corn-mill, in full view of Willington viaduct, on the Newcastle and Shields Railway; and it may not be irrelevant to mention that it is quite detached from the mill, or any other premises, and has no cellaring under it. The proprietor of the house, who lives in it, declines to make public the particulars of the disturbance to which he has been subjected, and it must be understood that the account of the visit we are about to lay before our readers is derived from a friend to whom Dr. Drury presented a copy of his correspondence on the subject, with power to make such use of it as he thought proper. We learned that the house had been reputed, at least one room in it, to have been haunted forty years ago, and had afterwards been undisturbed for a long period, during some years of which quietude the present occupant lived in it unmolested. We are also informed, that about the time that the premises were building, viz., in 1800 or 1801, there were reports of some deed of darkness having been committed by some one employed about them. We should

extend this account beyond the limits we have
set to ourselves, did we now enter upon a full
account of the strange things which have been
seen and heard about the place by several of
the neighbours, as well as those which are reported to have been seen, heard, and felt, by
the inmates, whose servants have been
changed, on that account, many times. We
proceed, therefore, to give the following letters
which have been passed between individuals
of undoubted veracity; leaving the reader to
draw his own conclusions on the subject.

" (COPY, No, 1.)
" To Mr. Procter,   17th June, 1840.
" SIR,—Having heard from indisputable
authority, viz., that of my excellent friend,
Mr. Davison, of Low Willington, farmer, that
you and your family are disturbed by most unaccountable noises at night, I beg leave to tell
you that I have read attentively Wesley's
account of such things, but with, I must confess, no great belief; but an account of this
report coming from one of your sect, which I
admire for candour and simplicity, my curiosity
is excited to a high pitch, which I would fain
satisfy. My desire is to remain alone in the
house all night with no companion but my own

watch-dog, in which, as far as courage and
fidelity are concerned, I place much more reliance than upon any three young gentlemen I
know of. And it is also my hope, that, if I
have a fair trial, I shall be able to unravel this
mystery. Mr. Davison will give you every
satisfaction if you take the trouble to enquire
of him concerning me.

    " I am, Sir,
     " Yours most respectfully,
       " EDWARD DRURY.
" At C. C. Embleton's, Surgeon,
 " No. 10, Church Street, Sunderland.

    " (COPY, No. 2.)
 " Joseph Procter's respects to Edward Drury,
whose note he received a few days ago, expressing a wish to pass a night in his house,
at Willington. As the family is going from
home on the 23rd instant, and one of
Unthank and Procter's men will sleep in the
house, if E. D. feel inclined to come on or
after the 24th to spend a night in it, he is at
liberty so to do, with or without his faithful
dog, which, by the bye, can be of no possible
use, except as company. At the same time,
J. P. thinks it best to inform him, that particular disturbances are far from frequent at

present, being only occasional, and quite uncertain, and therefore the satisfaction of E. D.'s
curiosity must be considered as problematical.
The best chance will be afforded by his sitting
up alone in the third story, till it be fairly
daylight, say two or three, A.M.

" Willington, 6th mo. 21st, 1840.

" J. P. will leave word with T. Maun, foreman, to admit E. D.

" Mr. Procter left home with his family on
the 23rd of June, and got an old servant, who
was then out of place in consequence of ill-
health, to take charge of the house during
their absence.   Mr. P. returned alone, on
account of business, on the 3rd of July, on the
evening of which day Mr. Drury and his companion also unexpectedly arrived.   After the
house had been locked up, every corner of it
was minutely examined.   The room out of
which the apparition issued is too shallow to
contain any person.   Mr. Drury and his friend
had lights by them, and were satisfied that
there was no one in the house besides Mr. P.,
the servant, and themselves.

" (COPY, No. 3.)

" Monday Morning, July 6, 1840.

" To Mr. Procter.

" DEAR SIR, — I am sorry I was not

at home to receive you yesterday, when you
kindly called to inquire for me. I am happy to
state that I am really surprised that I have
been so little affected as I am, after that
horrid and most awful affair. The only bad
effect that I feel is a heavy dullness in one of
my ears, the right one. I call it heavy dullness,
because I not only do not hear distinctly, but
feel in it a constant noise. This I never was
affected with before; but I doubt not it will
go off. I am persuaded that no one went to
your house at any time more *disbelieoing in
respect to seeing anything peculiar;* now no
one can be more satisfied than myself. I
will, in the course of a few days, send you a full
detail of all I saw and heard. Mr. Spence and
two other gentlemen came down to my house
in the afternoon, to hear my detail; but, sir,
could I account for these noises from natural
causes, yet, so firmly am I persuaded of the
horrid apparition, that I would affirm that
what I saw with my eyes was a punishment
to me for my scoffing and unbelief; that I am
assured that, as far as the horror is concerned,
they are happy that believe and have not seen.
Let me trouble you, sir, to give me the address
of your sister, from Cumberland, who was

alarmed, and also of your brother. I would
feel a satisfaction in having a line from them;
and, above all things, it will be a great cause
of joy to me, if you never allow your young
family to be in that horrid house again.
Hoping you will write a few lines at your
leisure,

" I remain, dear Sir,

" Yours very truly,

" EDWARD DRURY.

" (COPY, No. 4.)

" Willington, 7th mo. 9, 1840.

" Respected Friend, E. Drury,

" Having been at Sunderland, I did not
receive thine of the 6th till yesterday morning.
I am glad to hear thou art getting well over
the effects of thy unlooked-for visitation. I
hold in respect thy bold and manly assertion
of the truth in the face of that ridicule and
ignorant conceit with which that which is
called the supernatural, in the present day, is
usually assailed.

" I shall be glad to receive thy detail, in
which it will be needful to be very particular
in showing that thou couldst not be asleep, or

attacked by nightmare, or mistake a reflection of the candle, as some sagaciously suppose.

"I remain, respectfully,

"Thy friend,

"JOSH. PROCTER.

"P.S.—I have about thirty witnesses to various things which cannot be satisfactorily accounted for on any other principle than that of spiritual agency.

"(COPY, No. 5.)

"Sunderland, July 13, 1840.

"DEAR SIR,—I hereby, according to promise in my last letter, forward you a true account of what I heard and saw at your house, in which I was led to pass the night from various rumours circulated by most respectable parties, particularly from an account by my esteemed friend Mr. Davison, whose name I mentioned to you in a former letter. Having received your sanction to visit your mysterious dwelling, I went on the 3rd of July, accompanied by a friend of mine, T. Hudson. This was not according to promise, nor in accordance with my first intent, as I wrote you I would come alone; but I felt gratified at your kindness in not alluding to the liberty I had

VOL. II.   N

taken, as it ultimately proved for the best. I
must here mention that, not expecting you
at home, I had in my pocket a brace of
pistols, determining in my mind to let one of
them drop before the miller, as if by accident,
for fear he should presume to play tricks upon
me; but after my interview with you, I felt
there was no occasion for weapons, and did
not load them, after you had allowed us to
inspect as minutely as we pleased every portion of the house. I sat down on the third
story landing, fully expecting to account for
any noises that I might hear, in a philosophical
manner. This was about eleven o'clock, P.M.
About ten minutes to twelve we both heard a
noise, as if a number of people were pattering
with their bare feet upon the floor; and yet, so
singular was the noise, that I could not minutely determine from whence it proceeded.
A few minutes afterwards we heard a noise,
as if some one was knocking with his knuckles
among our feet; this was followed by a hollow
cough from the very room from which the
apparition proceeded. The only noise after
this, was as if a person was rustling against the
wall in coming up stairs. At a quarter to one,
I told my friend that, feeling a little cold, I
would like to go to bed, as we might hear the

noise equally well there; he replied that he would not go to bed till daylight. I took up a note which I had accidentally dropped, and began to read it, after which I took out my watch to ascertain the time, and found that it wanted ten minutes to one. In taking my eyes from the watch, they became rivetted upon a closet door, which I distinctly saw open, and saw also the figure of a female attired in grayish garments, with the head inclining downwards, and one hand pressed upon the chest, as if in pain, and the other, viz , the right-hand, extended towards the floor, with the index finger pointing downwards. It advanced with an apparently cautious step across the floor towards me; immediately as it approached my friend, who was slumbering, its right hand was extended towards him; I then rushed at it, giving, as Mr. Procter states, a most awful yell; but, instead of grasping it, I fell upon my friend, and I recollected nothing distinctly for nearly three hours afterwards. I have since learnt that I was carried down stairs in an agony of fear and terror.

" I hereby certify that the above account is strictly true and correct in every respect.

" North Shields. " EDWARD DRURY.

"The following more recent case of an apparition seen in the window of the same house
from the outside, by four credible witnesses,
who had the opportunity of scrutinizing it for
more than ten minutes, is given on most un-
questionable authority. One of these witnesses
is a young lady, a near connexion of the family,
who, for obvious reasons, did not sleep in the
house; another, a respectable man, who has
been many years employed in, and is foreman
of, the manufactory; his daughter, aged about
seventeen; and his wife who first saw the object,
and called out the others to view it. The
appearance presented was that of a bareheaded
man, in a flowing robe like a surplice, who
glided backwards and forwards about three
feet from the floor, or level with the bottom of
the second story window, seeming to enter the
wall on each side, and thus present a side view
in passing. It then stood still in the window,
and a part of the body came through both the
blind, which was close down, and the window,
as its luminous body intercepted the view of
the framework of the window. It was semi-
transparent, and as bright as a star, diffusing
a radiance all around. As it grew more dim,
it assumed a blue tinge, and gradually faded
away from the head downwards. The fore-

man passed twice close to the house under the window, and also went to inform the family, but found the house locked up. There was no moonlight, nor a ray of light visible anywhere about, and no person near. Had any magic lantern been used, it could not possibly have escaped detection; and it is obvious nothing of that kind could have been employed on the inside, as in that case the light could only have been thrown upon the blind, and not so as to intercept the view both of the blind and of the window from without. The owner of the house slept in that room, and must have entered it shortly after this figure had disappeared.

" It may well be supposed what a sensation the report of the visit of Mr. Drury, and its result must have created. It flew far and wide, and when it appeared in print, still wider; and what was not a little singular, Mr. Procter received, in consequence, a great number of letters, from individuals of different ranks and circumstances, including many of much property, informing him that their residences were, and had been for years, subject to annoyances of precisely a similar character

" So the ghosts and the hauntings are not gone, after all! We have turned our backs on them, and, in the pride of our philosophy, have refused to believe in them ; but they have persisted in remaining, notwithstanding !

"These singular circumstances being at various times related by parties acquainted with the family at Willington, I was curious, on a tour northward some time ago, to pay this haunted house a visit, and to solicit a night's lodgings there. Unfortunately the family was absent, on a visit to Mrs. Procter's relatives in Carlisle, so that my principal purpose was defeated; but I found the foreman and his wife, mentioned in the foregoing narrative, living just by. They spoke of the facts above detailed with the simple earnestness of people who had no doubts whatever on the subject. The noises and apparitions in and about this house seemed just like any other facts connected with it—as matters too palpable and positive to be questioned, any more than that the house actually stood, and the mill ground. They mentioned to me the circumstance of the young lady, as above stated, who took up her lodging in their house, because she would no longer encounter the annoyances of the haunted house ; and what trouble it had occasioned the family in procuring and retaining servants.

"The wife accompanied me into the house,

which I found in charge of a recently married servant and her husband, during the absence of the family. This young woman, who had, previous to her marriage, lived some time in the house, had never seen anything, and therefore had no fear. I was shown over the house, and especially into the room on the third story, the main haunt of the unwelcome visitors, and where Dr. Drury had received such an alarm. This room, as stated, was, and had been for some time, abandoned as a bed-room, from its bad character, and was occupied as a lumber-room.

"At Carlisle, I again missed Mr. Procter; he had returned to Willington, so that I lost the opportunity of hearing from him or Mrs. Procter, any account of these singular matters. I saw, however, various members of his wife's family, most intelligent people, of the highest character for sound and practical sense, and they were unanimous in their confirmation of the particulars I had heard, and which are here related.

"One of Mrs. Procter's brothers, a gentleman in middle life, and of a peculiarly sensible, sedate, and candid disposition, a person apparently most unlikely to be imposed on by fictitious alarms or tricks, assured me me that he had himself, on a visit there, been disturbed by the strangest noises. That he had resolved, before going, that if any such noises occurred he would speak, and demand of the invisible actor who he was, and why he came thither. But the occasion came, and he found himself unable to fulfil his intention. As he lay in bed one night, he heard a heavy step ascend the stairs towards his room, and some one striking, as it were, with a thick stick on the banisters, as he went along. It came to his door, and he essayed to call, but his voice died in his throat. He then sprang from his bed, and opening the door, found no one there, but now heard the same heavy steps deliberately descending, though perfectly invisibly, the steps before his face, and accompanying the descent with the same loud blows on the banisters.

" My informant now proceeded to the room door of Mr. Procter, who, he found, had also heard the sounds, and who now also arose, and with a light they made a speedy descent below, and a thorough search there, but without discovering anything that could account for the occurrence.

" The two young ladies, who, on a visit there, had also been annoyed by this invisible agent, gave me this account of it :—The first night, as they were sleeping in the same bed, they felt the bed lifted up beneath them. Of course,

they were much alarmed. They feared lest
some one had concealed himself there for the
purpose of robbery. They gave an alarm,
search was made, but nothing was found. On
another night, their bed was violently shaken,
and the curtains suddenly hoisted up all round
to the very tester, as if pulled by cords, and
as rapidly let down again, several times.*
Search again produced no evidence of the
cause. The next, they had the curtains totally
removed from the bed, resolving to sleep without them, as they felt as though evil eyes were
lurking behind them. The consequences of
this, however, was still more striking and terrific. The following night, as they happened
to awake, and the chamber was light enough—
for it was summer—to see everything in it,
they both saw a female figure, of a misty substance, and bluish grey hue, come out of the
wall, at the bed's head, and through the head-
board, in a horizontal position, and lean over
them. They saw it most distinctly. They saw
it as a female figure come out of, and again
pass into, the wall. Their terror became intense, and one of the sisters, from that night,
refused to sleep any more in the house, but

* It is remarkable that this hoisting of the bed-curtains
is similar to an incident recorded in the account of the visit
of Lord Tyrone's ghost to Lady Beresford.

took refuge in the house of the foreman during her stay ; the other shifting her quarters to another part of the house. It was the young lady who slept at the foreman's who saw, as above related, the singular apparition of the luminous figure in the window, along with the foreman and his wife.

" It would be too long to relate all the forms in which this nocturnal disturbance is said by the family to present itself. When a figure appears, it is sometimes that of a man, as already described, which is often very luminous, and passes through the walls as though they were nothing. This male figure is well known to the neighbours by the name of " Old Jeffrey !" At other times, it is the figure of a lady also in gray costume, and as described by Mr. Drury. She is sometimes seen sitting wrapped in a sort of mantle, with her head depressed, and her hands crossed on her lap. The most terrible fact is that she is without eyes.

" To hear such sober and superior people gravely relate to you such things, gives you a very odd feeling. They say that the noise made is often like that of a paviour with his rammer thumping on the floor. At other times it is coming down the stairs, making a similar loud sound. At others it coughs, sighs, and groans,

like a person in distress; and, again, there is
the sound of a number of little feet pattering
on the floor of the upper chamber, where the
apparition has more particularly exhibited
itself, and which, for that reason, is solely
used as a lumber room. Here these little footsteps may be often heard as if careering a
child's carriage about, which in bad weather
is kept up there. Sometimes, again, it makes
the most horrible laughs. Nor does it always
confine itself to the night. On one occasion, a
young lady, as she assured me herself, opened
the door in answer to a knock, the housemaid
being absent, and a lady in fawn-coloured silk
entered, and proceeded up stairs. As the
young lady, of course, supposed it a neighbour
come to make a morning call on Mrs. Procter,
she followed her up to the drawing-room,
where, however, to her astonishment, she did
not find her, nor was anything more seen
of her.

"Such are a few of the 'questionable
shapes' in which this troublesome guest comes.
As may be expected, the terror of it is felt by
the neighbouring cottagers, though it seems to
confine its malicious disturbance almost solely
to the occupants of this one house. There is a

well, however, near to which no one ventures
after it is dark, because it has been seen
near it.

"It is useless to attempt to give any opinion
respecting the real causes of these strange
sounds and sights. How far they may be real
or imaginary, how far they may be explicable
by natural causes or not; the only thing which
we have here to record, is the very singular
fact of a most respectable and intelligent
family having for many years been continually
annoyed by them, as well as their visitors.
They express themselves as most anxious to
obtain any clue to the true cause, as may be
seen by Mr. Procter's ready acquiescence in
the experiment of Mr. Drury. So great a
trouble is it to them, that they have contem-
plated the necessity of quitting the house
altogether, though it would create great inconvenience as regarded business. And it only
remains to be added, that we have not heard
very recently whether these visitations are still
continued, though we have a letter of Mr.
Procter's to a friend of ours, dated September
1844, in which he says, ' Disturbances have
for a length of time been only very unfrequent, which is a comfort, as the elder children

are getting old enough (about nine or ten
years) to be more injuriously affected by anything of the sort.'

" Over these facts let the philosophers
ponder, and if any of them be powerful enough
to exorcise " Old Jeffery," or the bluish-grey
and misty lady, we are sure that Mr. Joseph
Procter will hold himself deeply indebted to
them. We have lately heard that Mr. Procter
has discovered an old book, which makes it
appear that the very same ' hauntings' took
place in an old house, on the very same spot,
at least two hundred years ago."

To the above information, furnished by
Mr. Howitt, I have to subjoin that the family
of Mr. Procter are now quitting the house,
which he intends to divide into small tene-
ments for the work-people. A friend of mine
who lately visited Willington, and who went
over the house with Mr. Procter, assures me
that the annoyances still continue, though less
frequent than formerly. Mr. P. informed her
that the female figure generally appeared in a
shroud, and that it had been seen in that guise
by one of the family only a few days before.
A wish being expressed by a gentleman
visiting Mr. P. that some natural explanation

of these perplexing circumstances might be discovered, the latter declared his entire conviction, founded on an experience of fifteen years, that no such elucidation was possible.

CHAPTER IV.

SPECTRAL LIGHTS, AND APPARITIONS ATTACHED
TO CERTAIN FAMILIES.

In commencing another chapter, I take the
opportunity of repeating what I have said before, viz., that in treating of these phenomena,
I find it most convenient to assume what I
myself believe, that they are to be explained
by the existence and appearance of what are
called *ghosts;* but in so doing, I am not presuming to settle the question : if any one will
examine into the facts and furnish a better
explanation of them, I shall be ready to receive it.

In the mean time, assuming this hypothesis, there is one phenomenon frequently attending their appearance, which has given rise to a great deal of thoughtless ridicule, but which, in the present state of science, merits very particular attention. Grose, whom Dr. Hibbert quotes with particular satisfaction, says, "I cannot learn that ghosts carry tapers in their hands, as they are sometimes depicted, though the room in which they appear, even when without fire or candle, is frequently said to be as light as day."

Most persons will have heard of this peculiarity attending the appearance of ghosts. In the case of Professor Dorrien's apparition, mentioned in a former chapter, Professor Oeder saw it, when there was no light in the room, by a flame which proceeded from itself. When he had the room lighted, he saw it no longer; the light of the lamp rendering invisible the more delicate phosphorescent light of the spectre; just as the bright glare of the sun veils the feebler lustre of the stars, and obscures to our senses many chemical lights, which are very perceptible in darkness. Hence the notion, so available to those who satisfy themselves with scoffing without enquiring, that broad daylight banishes apparitions, and

that the belief in them is merely the offspring
of physical as well as moral darkness.

I meet with innumerable cases in which
this phosphorescent light is one of the accompaniments, the flame sometimes proceeding
visibly from the figure; whilst in others, the
room appeared pervaded with light without its
seeming to issue from any particular object.

I remember a case of the servants in a
country house, in Aberdeenshire, hearing the
door-bell ring after their mistress was gone to
bed; on coming up to open it, they saw through
a window that looked into the hall, that it was
quite light, and that their master, Mr. F., who
was at the time absent from home, was there
in his travelling dress. They ran to tell their
mistress what they had seen ; but when they
returned, all was dark, and there was nothing
unusual to be discovered. That night Mr. F.
died at sea, on his voyage to London.

A gentleman, some time ago, awoke in the
middle of a dark winter's night, and perceived
that his room was as light as if it were day.
He awoke his wife and mentioned the circumstance, saying he could not help apprehending
that some misfortune had occurred to his fish-
ing boats, which had put to sea. The boats
were lost that night.

Only last year, there was a very curious circumstance happened in the south of England,
in which these lights were seen. I give the
account, literally, as I extracted it from the
newspaper, and also the answer of the editor
to my further enquiries. I know nothing more
of this story; but it is singularly in keeping
with others proceeding from different quarters.

"A GHOST AT BRISTOL.—We have this week
a ghost story to relate. Yes, a ghost story; a
real ghost story, and a ghost story without, as
yet, any clue to its elucidation. After the
dissolution of the Calendars, their ancient residence, adjoining, and almost forming a part
of All Saints' Church, Bristol, was converted
into a vicarage-house, and it is still called by
that name, though the incumbents have for
many years ceased to reside there. The present occupants are Mr. and Mrs. Jones, the
sexton and sextoness of the church, and one
or two lodgers; and it is to the former and
their servant-maid, that the strange visitor has
made his appearance, causing such terror by
his nightly calls, that all three have determined
on quitting the premises, if, indeed, they have
not already carried their resolution into effect.
Mr. and Mrs. Jones's description of the disturbance, as given to the landlord, on whom
they called in great consternation, is as distinct

as any ghost story could be. The nocturnal
visitor is heard walking about the house when
the inhabitants are in bed; and Mr. Jones, who
is a man of by no means nervous constitution, declares he has several times seen a light
flickering on one of the walls. Mrs. Jones is
equally certain that she has heard a man with
creaking shoes walking in the bed-room above
her own, when no man was on the premises
(or at least ought to be), and " was nearly
killed with the fright." To the servant maid,
however, was vouchsafed the unenvied honour
of seeing this restless night visitor; she declares
she has repeatedly had her bed-room door
unbolted at night, between the hours of twelve
and two o'clock—the period when such beings
usually make their promenades—by something
in human semblance; she cannot particularize
his dress, but describes it as something antique,
and of a fashion "lang syne gane," and to
some extent corresponding to that of the
ancient Calendars, the former inhabitants of
the house. She further says, he is a " whiskered gentleman" (we give her own words),
which whiskered gentleman has gone the
length of shaking her bed, and, she believes,
would have shaken herself also, but that she
invariably puts her head under the clothes

when she sees him approach. Mrs. Jones declares she believes in the appearance of the whiskered gentleman, and she had made up her mind, the night before she called on her landlord, to leap out of the window (and it is not a trifle that will make people leap out of windows), as soon as he entered the room. The effect of the 'flickering light' on Mr. Jones was quite terrific, causing excessive trembling, and the complete doubling up of his whole body into a round ball, like."— *Bristol Times.*

" Bristol Times Office,
3rd June, 1846.

" Madam, — In reply to your enquiries respecting the ghost story, I have to assure you that the whole affair remains wrapped in the same mystery as when chronicled in the pages of the *Bristol Times.*

" I am, Madam,
"Yours obediently,
" The Editor."

I subsequently wrote to Mrs. Jones, who I found was not a very dexterous scribe, but she confirmed the above account, adding, however, that the Rev. Mr. —, the clergyman of the parish, said I had better write to him about it, and that he does not believe in such things."

Of course, he does not; and it would have been useless to have asked his opinion.

There never was, perhaps, a more fearless human being than Madame Gottfried, the Empoisonneuse of Bremen ; at least, she felt no remorse—she feared nothing but discovery ; and yet, when after years of successful crime, she was at length arrested, she related, that soon after the death of her first husband, Miltenburg, whom she had poisoned, as she was standing, in the dusk of the evening, in her drawing-room, she suddenly saw a bright light hovering at no great distance above the floor, which advanced towards her bed-room door and then disappeared. This phenomenon occurred on three successive evenings. On another occasion, she saw a shadowy appearance hovering near her, " Ach ! denke ich, das. ist Miltenburg, seine Erscheinung !—Alas ! thought I, that is the ghost of Miltenburg ! " Yet did not this withhold her murderous hand.

The lady who met with the curious adventure in Petersburgh, mentioned in a former chapter, had no light in her room ; yet she saw the watch distinctly by the old woman's light, though of what nature it was, she does not know. Of the lights seen over graves, familiarly called *corpse candles*, I have spoken

elsewhere; as also of the luminous form perceived by Rilling, in the garden at Colmar, as
mentioned by Baron von Reichenbach.   Most
people have heard the story of the Radiant
Boy, seen by Lord Castlereagh, an apparition
which the owner of the castle, admitted to
have been visible to many others.   Dr. Kerner
mentions a very similar fact, wherein an advocate and his wife were awakened by a noise
and a light, and saw a beautiful child enveloped by the sort of glory that is seen surrounding the heads of saints.   It disappeared,
and they never had a repetition of the phenomenon, which they afterwards heard was believed to recur every seven years in that house,
and to be connected with the cruel murder of
a child by its mother.

To these instances I will add an account of
the ghost seen in C— Castle, copied from the
handwriting of C. M. H., in a book of manu-
script extracts; dated C—Castle, December
22nd, 1824 ; and furnished to me by a friend
of the family.

" In order to introduce my readers to the
haunted room, I will mention that it forms
part of the old house, with windows looking
into the court, which in early times was deemed
a necessary security against an enemy.   It ad-

joins a tower built by the Romans for defence ;
for C— was properly more a border tower than
a castle of any consideration. There is a wind-
ing staircase in this tower, and the walls are
from eight to ten feet thick.

" When the times became more peaceable,
our ancestors enlarged the arrow-slit windows,
and added to that part of the building which
looks towards the river Eden ; the view of
which, with its beautiful banks, we now enjoy·
But many additions and alterations have been
made since that.

" To return to the room in question, I must
observe that it is by no means remote or solitary, being surrounded on all sides by chambers
that are constantly inhabited. It is accessible
by a passage cut through a wall eight feet in
thickness, and its dimensions are twenty-one
by eighteen. One side of the wainscoating is
covered with tapestry, the remainder is deco-
rated with old family pictures, and some
ancient pieces of embroidery, probably the
handiwork of nuns. Over a press, which has
doors of Venetian glass, is an ancient oaken
figure, with a battle-axe in his hand, which
was one of those formerly placed on the walls
of the city of Carlisle, to represent guards.
There used to be, also, an old-fashioned bed

and some dark furniture in this room ; but, so many were the complaints of those who slept there, that I was induced to replace some of these articles of furniture by more modern ones, in the hope of removing a certain air of glooom, which I thought might have given rise to the unaccountable reports of apparitions and extraordinary noises which were constantly reaching us. But I regret to say I did not succeed in banishing the nocturnal visitor, which still continues to disturb our friends.

" I shall pass over numerous instances, and select one as being especially remarkable, from the circumstance of the apparition having been seen by a clergyman well known and highly respected in this county, who, not six weeks ago, repeated the circumstances to a company of twenty persons, amongst whom were some who had previously been entire disbelievers in such appearances.

" The best way of giving you these particulars, will be by subjoining an extract from my journal, entered at the time the event occurred.

" SEPT. 8, 1803. — Amongst other guests invited to C— Castle, came the Rev. Henry A., of Redburgh, and rector of Greystoke, with

Mrs. A. his wife, who was a Miss S., of Ulverstone. According to previous arrangements, they were to have remained with us some days; but their visit was cut short in a very unexpected manner. On the morning after their arrival, we were all assembled at breakfast, when a chaise and four dashed up to the door in such haste, that it knocked down part of the fence of my flower-garden. Our curiosity was, of course, awakened to know who could be arriving at so early an hour ; when, happening to turn my eyes towards Mr. A., I observed that he appeared extremely agitated. ' It is our carriage !' said he ; ' I am very sorry, but we must absolutely leave you this morning.'

" We naturally felt and expressed considerable surprise, as well as regret, at this unexpected departure ; representing that we had invited Colonel and Mrs. S., some friends whom Mr. A. particularly desired to meet, to dine with us on that day. Our expostulations however were vain ; the breakfast was no sooner over than they departed, leaving us in consternation to conjecture what could possibly have occasioned so sudden an alteration in their arrangements. I really felt quite uneasy lest anything should have given them offence ; and we reviewed all the occurrences of the preceding

evening in order to discover, if offence there
was, whence it had arisen. But our pains were
vain; and after talking a great deal about it
for some days, other circumstances banished
it from our minds.

"It was not till we some time afterwards
visited the part of the county in which Mr. A.
resides, that we learnt the real cause of his
sudden departure from C—. The relation of
the fact, as it here follows, is in his own
words:—

"Soon after we went to bed, we fell asleep:
it might be between one and two in the morning when I awoke. I observed that the fire
was totally extinguished; but although that
was the case, and we had no light, I saw a
glimmer in the centre of the room, which suddenly increased to a bright flame. I looked
out, apprehending that something had caught
fire; when, to my amazement, I beheld a
beautiful boy, clothed in white, with bright
locks, resembling gold, standing by my bed-
side, in which position he remained some
minutes, fixing his eyes upon me with a mild
and benevolent expression. He then glided
gently away towards the side of the chimney,
where it is obvious there is no possible egress;
and entirely disappeared. I found myself

again in total darkness, and all remained quiet until the usual hour of rising. I declare this to be a true account of what I saw at C— Castle, upon my word as a clergyman."

I am acquainted with some of the family, and with several of the friends of Mr. A., who is still alive, though now an old man; and I can most positively assert that his own conviction, with regard to the nature of this appearance, has remained ever unshaken. The circumstance made a lasting impression upon his mind, and he never willingly speaks of it; but when he does, it is always with the greatest seriousness, and he never shrinks from avowing his belief, that what he saw admits of no other interpretation than the one he then put upon it.

Now, let us see whether in this department of the phenomenon of ghost-seeing, namely, the lights that frequently accompany the apparitions, there is anything so worthy of ridicule as Grose, and other such commentators seem to think.

Of God, the uncreated, we know nothing; but the created spirit, man, we cannot conceive of independant of some organism or organ, however different that organ may be to those which form our means of apprehension and communication at present. This organ, we may suppose to be that pervading ether, which is now the germ of what St. Paul calls our *spiritual body*, the *astral spirit* of the mystics, he *nerve-spirit* of the clear-seers; the fundamental body, of which the external fleshly body is but the copy and husk—an organ comprehending all those distinct ones, which we now possess in the one universal, or, as some of the German physiologists call it, the *central* sense, of which we occasionally obtain some glimpses in somnambulism, and in other peculiar states of nervous derangement; especially where the ordinary senses of sight, hearing, feeling, &c. are in abeyance; an effect which Dr. Ennemoser considers to be produced by a change of polarity, the external periphery of the nerves taking on a negative state; and which Dr. Passavent describes as the retreating of the ether from the external to the internal, so that the nerves no longer receive impressions, or convey information to the brain; a condition which may be produced by various causes, as excess of excitement, great elevation of the spirit, as we see in the extatics and martyrs, over irritation producing consequent exhaustion; and also artificially, by certain manipulations, narcotics, and other influences. All somnambules of the highest order—and when I make use of this expression, I repeat that I do not allude to the subjects of mesmeric experiments, but to those extraordinary cases of disease, the particulars of which have been recorded by various continental physicians of eminence—all persons in that condition describe themselves as hearing and seeing, not by their ordinary organs, but by some means, the idea of which they cannot convey further than that they are pervaded by light, and that this is not the *ordinary* physical light is evident, inasmuch as that they generally see best in the dark, a remarkable instance of which I myself witnessed.

I never had the slightest idea of this internal light, till in the way of experiment, I inhaled the sulphuric ether; but I am now very well able to conceive it; for after first feeling an agreeable warmth pervading my limbs, my next sensation was to find myself, I cannot say in this heavenly light, for the light was in *me*—I was pervaded by it; it was not perceived by my eyes which were closed, but perceived internally, I cannot tell how. Of what nature this heavenly light was, and I cannot forbear calling it *heavenly*, for it was like nothing on earth—I know not, nor how far it may be related to those lumi-

nous emanations occasionally seen around
extatics, saints, martyrs, and dying persons;
or to the flames seen by somnambules issuing
from various objects, or to those observed by
Von Reichenbach's patients proceeding from
the ends of the fingers, &c. But at all events,
since the process which maintains life is of
the nature of combustion, we have no reason
to be amazed at the presence of luminous
emanations; and as we are the subjects of
various electrical phenomena, nobody is surprised when, on combing their hair or pulling
off their silk stockings, they hear a crackling
noise or even see sparks.

Light, in short, is a phenomenon which
seems connected with all forms of life ; and I
need not here refer to that emitted by glow-
worms, fire-flies and those marine animals,
which illuminate the sea. The eyes also of
many animals shine with a light which is not
merely a reflected one; as has been ascertained by Rengger, a German naturalist, who
found himself able to distinguish objects in
the most profound darkness, by the flaming
eyes of a South American monkey.

"The seeing of a clear-seer," says Dr.
Passavent, " may be called a *solar* seeing, for
he lights and inter-penetrates his object with

his *own* organic light, viz., his nervous ether, which becomes the organ of the spirit ; and under certain circumstances this organic light becomes visible, as in those above alluded to. Persons recovering from deep swoons and trances, frequently describe themselves as having been in this region of light—this light of the spirit, if I may so call it—this palace of light, in which it dwells, which will hereafter be its proper light, for the physical or solar light, which serves us whilst in the flesh, will be no longer needed, when out of it, nor probably be perceived by the spirit, which will then, I repeat, be a light to itself; and as this organic light, this germ of our future spiritual body, occasionally becomes partially visible now, there cannot, I think, be any great difficulty in conceiving, that it may under given circumstances, be so hereafter.

The use of the word *light* in the scriptures, must not be received in a purely symbolical sense. We shall dwell in light, or we shall dwell in darkness, in proportion as we have shaken off the bonds that chain us to the earth; according, in short, to our moral state, we shall be pure and bright, or impure and dark.

Monsieur Arago mentions in his treatise on lightning and the electrical fluid, that all

men are not equally susceptible of it; and
that, there are different degrees of receptivity,
verging from total insensibility to the extreme
opposite ; and he also remarks, that animals
are more susceptible to it than men. He says,
the fluid will pass through a chain of persons,
of whom, perhaps one ; though forming only
the second link, will be wholly insensible of
the shock.   Such persons would be rarely
struck by lightning, whilst another would
be in as great danger from a flash, as if he
were made of metal.   Thus it is not only the
situation of a man, during a storm, but also
his physical constitution, that regulates the
amount of his peril.   The horse and the dog
are particularly susceptible.

Now, this varying susceptibility, is analogous
to, if not the very same, that causes the varying
susceptibility to such phenomena as I am
treating of; and, accordingly, we know that in
all times, horses and dogs have been reputed
to have the faculty of seeing spirits; and when
persons who have the second sight see a
vision, these animals, if in contact with them,
perceive it also, and frequently evince symptoms of great terror. We also here find the
explanation of another mystery, namely, what
the Germans call *ansteckung,* and the English,

sceptics when alluding to these phenomena, *contagion*—meaning simply *contagious fear* ; but as when several persons form a chain, the shock from an electrical machine, will pass through the whole of them ; so if one person is in such a state as to become sensible of an apparition or some similar phenomenon, he may be able to communicate that power to another ; and thus has arisen the conviction amongst the Highlanders, that a seer by touching a person near him, enables him frequently to participate in his vision.

A little girl, in humble life, called Mary Delves, of a highly nervous temperament, has been frequently punished for saying that the cat was on fire ; and that she saw flames issuing from various persons and objects.

With regard to the perplexing subject of corpse lights, there would be little difficulty attending it, if they always remained stationary over the graves; but it seems very well established that that is not the case. There are numerous stories, proceding from very respectable quarters, proving the contrary ; and I have heard two from a dignitary of the church, born in Wales, which I will relate.

A female relation of his had occasion to go to Aberystwith, which was about twenty miles

from her home, on horseback; and she started
at a very early hour for that purpose, with her
father's servant.   When they had nearly
reached the half-way, fearing the man might be
wanted at home, she bade him return, as she
was approaching the spot where the servant
of the lady she was going to visit, was to meet
her, in order to escort her the other half. The
man had not long left her, when she saw a
light coming towards her, the nature of which
she suspected; it moved, according to her description, steadily on, about three feet from the
ground.   Somewhat awe-struck, she turned
her horse out of the bridle-road, along which
it was coming, intending to wait till it had
passed; but, to her dismay, just as it came
opposite to her, it stopped, and there remained
perfectly fixed for nearly half an hour ; at the
end of which period, it moved on as before.

The servant presently came up, and she
proceeded to the house of her friend, where
she related what she had seen.   A few days
afterwards, the very servant who came to meet
her, was taken ill and died; his body was
carried along that road; and at the very
spot where the light had paused, an accident
occurred, which caused a delay of half an hour.

The other story was as follows :—A servant

in the family of Lady Davis, my informant's
aunt, had occasion to start early for market.
Being in the kitchen, about three o'clock in
the morning, taking his breakfast alone, when
everybody else was in bed, he was surprised at
hearing a sound of heavy feet on the stairs
above; and opening the door to see who it
could be, he was struck with alarm at perceiving a great light, much brighter than could
have been shed by a candle, at the same time
that he heard a violent thump, as if some very
heavy body had hit the clock, which stood on
the landing. Aware of the nature of the light,
the man did not await its further descent, but
rushed out of the house; whence he presently
saw it issue from the front door, and proceed
on its way to the churchyard.

As his mistress, Lady D., was at that period
in her bed, ill, he made no doubt that her death
impended; and when he returned from the
market at night, his first question was whether
she was yet alive; and though he was informed
she was better, he declared his conviction that
she would die, alleging as his reason what he
had seen in the morning; a narration which
led everybody else to the same conclusion.

The lady however recovered; but within a
fortnight, another member of the family died;

and as his coffin was brought down the stairs,
the bearers ran it violently against the clock ;
upon which the man instantly exclaimed;
" That is the very noise I heard !"

I could relate numerous stories wherein the
appearance of a ghost was accompanied by a
light, but as there is nothing that distinguishes
them from those above-mentioned, I will not
dilate further on this branch of the subject,
on which, perhaps, I have said enough to suggest to the minds of my readers, that although
we know little *how* such things are, we do
know enough of analagous phenomena to enable us to believe, at least, their possibility.

I confess I find much less difficulty in conceiving the existence of such facts as those
above described, than those of another class, of
which we meet with occasional instances.

For example, a gentleman of fortune and
station, in Ireland, was one day walking along
the road, when he met a very old man, apparently a peasant, though well dressed, and
looking as if he had on his Sunday habiliments.
His great age attracted the gentleman's attention the more, that he could not help wondering at the alertness of his movements, and the
ease with which he was ascending the hill.
He consequently accosted him, enquiring his

name and residence; and was answered, that
his name was Kirkpatrick, and that he lived at
a cottage, which he pointed out. Whereupon
the gentleman expressed his surprise that he
should be unknown to him, since he fancied
he had been acquainted with every man on his
estate. "It is odd you have never seen me
before," returned the old man; "for I walk
here every day."

"How old are you!" asked the gentleman.

"I am one hundred and five," answered the
other; "and have been here all my life."

After a few more words, they parted; and the
gentleman proceeding towards some labourers
in a neighbouring field, enquired if they knew
an old man of the name of Kirkpatrick. They
did not; but on addressing the question to
some older tenants, they said, "Oh, yes;" they
had known him, and had been at his funeral;
he had lived at the cottage on the hill, but had
been dead twenty years.

"How old was he when he died?" enquired
the gentleman, much amazed. — "He was
eighty-five," said they; so that the old man
gave the age that he would have reached had
he survived to the period of this rencontre.

This curious incident is furnished by the gentleman himself, and all he can say is, that it

certainly occurred, and that he is quite unable to explain it. He was in perfect health at the time, and had never heard of this man in his life, who had been dead several years before the estate came into his possession.

The following is another curious story. The original will be found in the Register of the church named, from which it has been copied for my use :—

Extract from the Register in Brisley Church, Norfolk.

"December 12th, 1706.—I Robert Withers, M.A. vicar of Gately, do insert here a story which I had from undoubted hands; for I have all the moral certainty of the truth of it possible :—

"Mr Grose went to see Mr. Shaw on the 2nd of August last. As they sat talking in the evening, says Mr. Shaw 'On the 21st. of the last month, as I was smoking a pipe, and reading in my study, between eleven and twelve at night, in comes Mr. Naylor (formerly Fellow of St. John's College, but had been dead full four years) When I saw him I was not much affrighted, and I asked him to sit down, which accordingly he did for about two hours, and we talked together. I asked him how it fared with him? he said 'Very well.'

' Were any of our old acquaintance with him?'
' No!' (at which I was much concerned) 'but
Mr. Orchard will be with me soon and yourself not long after.' As he was going away I
asked him if he would not stay a little longer,
but he refused. I asked him if he would call
again. ' No;' he had but three days leave of
absence, and he had other business.

" N.B.—Mr. Orchard died soon after. Mr.
Shaw is now dead, he was formerly Fellow of
St. John's College, an ingenious good man.
I knew him there; but at his death he had a
college living in Oxfordshire, and here he saw
the apparition."

An extraordinary circumstance occurred
some years ago, in which a very pious and
very eminent Scotch minister, Ebenezer
Brown of Inverkeithing was concerned. A
person of ill character in the neighbourhood,
having died, the family very shortly afterwards came to him to complain, of some exceedingly unpleasant circumstances connected
with the room in which the dissolution had
taken place, which rendered it uninhabitable;
and requesting his assistance. All that is
known by his family of what followed is, that
he went and entered the room alone; came

out again; in a state of considerable excitement and in a great perspiration; took off his
coat, and re-entered the room; a great noise,
and, I believe voices were then heard by the
family, who remained during the whole time
at the door; when he came out finally, it was
evident that something very extraordinary had
taken place; what it was, he said, he could
never disclose; but that perhaps after his
death some paper might be found upon the
subject. None, however, as far I can learn,
has been discovered.

A circumstance of a very singular nature,
is asserted to have occurred, not very many
years back in regard to a professor in the
College of A—, who had seduced a girl and
afterwards married another woman. The girl
became troublesome to him, and being found
murdered after having been last seen in his
company, he was suspected of being some way
concerned in the crime; but the strange thing
is, that, from that period, he retired every
evening at a particular hour to a certain room,
where he staid great part of the night, and
where it was declared that *her* voice was distinctly heard in conversation with him, A
strange wild story, which I give as I have it,

without pretending to any explanation of the belief that seems to have prevailed, that he was obliged to keep this fearful tryst.

Visitations of this description, which seem to indicate that the deceased person is still, in some way incomprehensible to us, an inhabitant of the earth, are more perplexing than any of the stories I meet with. In the time of Frederick II. of Prussia, the cook of a Catholic priest, residing at a village named Quarrey, died, and he took another in her place; but the poor woman had no peace or rest from the interference of her predecessor, insomuch that she resigned her situation, and the minister might almost have done without any servant at all. The fires were lighted, and the rooms swept and arranged, and all the needful services performed, by unseen hands. Numbers of people went to witness the phenomena, till, at length, the story reached the ears of the King; who sent a captain and a lieutenant of his guard to investigate the affair. As they approached the house, they found themselves preceded by a march, though they could see no musicians, and when they entered the parlour and witnessed what was going on, the captain exclaimed, "If that doesn't beat the devil!" upon which he received a smart slap on the

Q 5

face, from the invisible hand that was arranging the furniture.

In consequence of this affair, the house was pulled down, by the King's orders, and another residence built for the minister, at some distance from the spot.

Now, to impose on Frederick II. would have been no slight matter, as regarded the probable consequences; and the officers of his guard would certainly not have been disposed to make the experiment; and it is not likely that the King would have ordered the house to be pulled down without being thoroughly satisfied of the truth of the story.

One of the most remarkable stories of this class I know, excepting indeed the famous one of the Grecian bride, is that which is said to have happened at Crossen, in Silesia, in the year 1659, in the reign of the Princess Elizabeth Charlotta. In the spring of that year, an apothecary's man, called Christopher Monig, a native of Serbest, in Anhalt, died, and was buried with the usual ceremonies of the Lutheran Church. But to the amazement of everybody; a few days afterwards, he, at least what appeared to be himself, appeared in the shop, where he would sit himself down, and sometimes walk, and take boxes, pots,

and glasses from the shelves, and set them
again in other places; sometimes trying and
examining the goodness of the medicines :
weigh them with the scales, pound the drugs
with a mighty noise; nay, serve the people
that came with bills to the shop, take their
money and lay it up in the counter; in
a word, do all things that a journeyman in
such cases used to do. He looked very ghostly
upon his former companions, who were afraid
to say anything to him, and his master being
sick at that time, he was very troublesome to
him. At last, he took a cloak that hung in
the shop, put it on and walked abroad, but
minding nobody in the streets ; he entered into
some of the citizen's houses, especially such as
he had formerly known, yet spoke to no one,
but to a maid servant, whom he met with hard
by the churchyard, whom he desired to go
home and dig in a lower chamber of her
master's house, where she would find an in-
estimable treasure. But the girl, amazed at
the sight of him, swooned away ; whereupon
he lifted her up, but left a mark upon her in so
doing, that was long visible. She fell sick in
consequence of the fright, and having told
what Monig had said to her, they dug up the
place indicated, but found nothing but a de-

cayed pot with an hemarites or bloodstone in
it. The affair making a great noise, the reigning Princess caused the man's body to be
taken up, which being done, it was found in a
state of putrefaction, and was re-interred.
The apothecary was then recommended to
remove everything belonging to Monig; his
linen, clothes, books, &c., after which, the
apparition left the house and was seen no more.

The fact of the man's re-appearance in this
manner, was considered to be so perfectly
established at the time, that there was actually
a public disputation on the subject, in the
Academy of Leipsick. With regard to the
importance the apparition attached to the
bloodstone, we do not know but that there
may be truth in the persuasion, that this gem
is possessed of some occult properties of much
more value than its beauty.

The story of the Grecian bride, is still more
wonderful, and yet it comes to us surprisingly
well authenticated, inasmuch as the details
were forwarded by the prefect of the city, in
which the thing occurred, to the pro-consul of
his province; and by the latter were laid before the Emperor Hadrian; and as it was not
the custom to mystify Roman Emperors, we
are constrained to believe that what the prefect

and pro-consul communicated to him, they had good reason for believing themselves.

It appears that a gentleman, called Demostratos, and Charito, his wife, had a daughter called Philinnion, who died; and that about six months afterwards, a youth named Machates, who had come to visit them, was surprised, on retiring to the apartments destined to strangers, by receiving the visits of a young maiden, who eats and drinks, and exchanges gifts with him. Some accident having taken the nurse that way, she, amazed by the sight, summons her master and mistress to behold their daughter, who is there sitting with the guest.

Of course, they do not believe her; but at length wearied by her importunities the mother follows her to the guest's chamber; but the young people are now asleep and the door closed; but looking through the keyhole, she perceives what she believes to be her daughter. Still unable to credit her senses, she resolves to wait till morning, before disturbing them; but when she comes again, the young lady has departed; whilst Machates, on being interrogated, confesses, that Philinnion had been with him, but that she had admitted to him that it was unknown

to her parents. Upon this, the amazement
and agitation of the mother were naturally
very great; especially when Machates showed
her a ring which the girl had given him, and
a bodice which she had left behind her; and
his amazement was no less, when he heard the
story they had to tell.   He however promised
that if she returned the next night, he would
let them see her ; for he found it impossible
to believe that his bride was their dead
daughter.   He suspected, on the contrary,
that some thieves had stripped her body of
the clothes and ornaments in which she had
been buried, and that the girl who came to
his room, had bought them.   When therefore
she arrived, his servant having had orders, to
summon the father and mother, they came;
and perceiving that it was really their
daughter they fell to embracing her, with
tears.   But she reproached them for the
intrusion, declaring that she had been permitted to spend three days with this stranger,
in the house of her birth ; but that now she
must go to the appointed place; and immediately fell down dead, and the dead body lay
there visible to all.

The news of this strange event soon spread
abroad, the house was surrounded by crowds

of people and the prefect was obliged to take measures to prevent a tumult. On the following morning at an early hour the inhabitants assembled in the theatre, and from thence they proceeded to the vault, in order to ascertain if the body of Philinnion was where it had been deposited six months before. It was not; but on the bier there lay the ring and cap which Machates had presented to her the first night she visited him; showing that she had returned there in the interim. They then proceeded to the house of Democrates, where they saw the body which it was decreed must now be buried without the bounds of the city. Numerous religious ceremonies and sacrifices followed, and the unfortunate Machates, seized with horror, put an end to his own life.

The following very singular circumstance occurred in this country towards the latter end of the last century; and excited, at the time, considerable attention; the more so, as it was asserted by everybody acquainted with the people and the locality, that the removal of the body was impossible, by any recognized means; besides that no one would have had the hardihood to attempt such a feat:—

"Mr. William Craighead, author of a popular

system of arithmetic, was parish schoolmaster
of Monifieth, situate upon the estuary of the
Tay, about six miles east from Dundee. It
would appear that Mr. Craighead was then a
young man, fond of a frolic, without being
very scrupulous about the means, or calculating
the consequences. There being a lykewake in
the neighbourhood, attended by a number of
his acquaintance, according to the custom of
the times; Craighead procured a confederate,
with whom he concerted a plan to draw the
watchers from the house, or at least from the
room where the corpse lay. Having succeeded
in this, he dexterously removed the dead body
to an outer house, while his companion occupied the place of the corpse in the bed where
it had lain. It was agreed upon between the
confederates, that when the company was re-
assembled, Craighead was to join them, and
at a concerted signal, the impostor was to rise,
shrouded like the dead man, while the two
were to enjoy the terror and alarm of their
companions. Mr. C. came in, and after being
some time seated, the signal was made, but
met no attention; he was rather surprised; it
was repeated, and still neglected. Mr. C., in
his turn, now became alarmed; for he con-
ceived it impossible that his companion could

have fallen asleep in that situation; his un-
easiness became insupportable; he went to
the bed, and found his friend lifeless!
Mr. C.'s feelings, as may well be imagined,
now entirely overpowered him, and the dreadful fact was disclosed; their agitation was
extreme, and it was far from being alleviated
when every attempt to restore animation to
the thoughtless young man proved abortive.
As soon as their confusion would permit, an
enquiry was made after the original corpse, and
Mr. C. and another went to fetch it in, but
it was not to be found. The alarm and consternation of the company was now redoubled;
for some time, a few suspected that some hardy
fellow among them had been attempting a
Rowland for an Oliver; but when every knowledge of it was most solemnly denied by all
present, their situation can be more easily
imagined than described; that of Mr. C. was
little short of distraction; daylight came without relieving their agitation; no trace of the
corpse could be discovered, and Mr. C. was
accused as the *primum mobile* of all that had
happened: he was incapable of sleeping, and
wandered several days and nights in search
of the body, which was at last discovered in
the parish of Tealing, deposited in a field about

six miles distant from the place from whence it was removed."

"It is related, that this extraordinary affair had a strong and lasting effect upon Mr. C.'s mind and conduct; that he immediately became serious and thoughtful, and ever after conducted himself with great prudence and sobriety."

Amongst what are called *superstitions*, there are a great many curious ones, attached to certain families; and from some members of these families, I have been assured, that experience has rendered it impossible for them to forbear attaching importance to these persuasions.

A very remarkable circumstance occurred lately in this part of the world, the facts of which I had an opportunity of being well acquainted with.

One evening, somewhere about Christmas, of the year 1844, a letter was sent for my perusal, which had been just received from a member of a distinguished family, in Perthshire. The friend who sent it me, an eminent literary man, said, " Read the enclosed ; and we shall now have an opportunity of observing if any event follows the prognostics." The information, contained in the letter, was to the following effect :—

Miss D. a relative of the present Lady C., who had been staying some time with the Earl and Countess, at their seat, near Dundee, was invited to spend a few days at C— Castle, with the Earl and Countess of A. She went: and whilst she was dressing for dinner, the first evening of her arrival, she heard a strain of music under her window, which finally resolved itself into a well defined sound of a drum. When her maid came up stairs, she made some enquiries about the drummer that was playing near the house; but the maid knew nothing on the subject. For the moment, the circumstance passed from Miss D.'s mind; but recurring to her again during the dinner, she said, addressing Lord A., "My Lord, who is your drummer?" upon which his lordship turned pale, Lady A. looked distressed, and several of the company, who all heard the question, embarrassed; whilst the lady, perceiving that she had made some unpleasant allusion, although she knew not to what their feelings referred, forebore further enquiry till she reached the drawing-room; when, having mentioned the circumstance again to a member of the family, she was answered, " What ! have you never heard of the drummer-boy !" " No ;" replied Miss D., " who in the world is he ?" " Why," replied the other, " he is a per-

son who goes about the house playing his
drum, whenever there is a death impending in
the family.   The last time he was heard, was
shortly before the death of the last Countess
(the Earl's former wife), and that is why Lord
A. became so pale when you mentioned it.
The drummer is a very unpleasant subject in
this family, I assure you!"

Miss D. was naturally much concerned and,
indeed, not a little frightened at this explanation, and her alarm being augmented by
hearing the sounds on the following day, she
took her departure from C— Castle and
returned to Lord C.'s, stopping on her way to
call on some friends, where she related this
strange circumstance to the family, through
whom the information reached me.

This affair was very generally known in the
north, and we awaited the event with interest.
The melancholy death of the Countess about
five or six months afterwards, at Brighton, sadly
verified the prognostic.   I have heard that a
paper was found in her desk after her death,
declaring her conviction that the dream was
for her; and it has been suggested that probably the thing preyed upon her mind and
caused the catastrophe; but in the first place,
from the mode of her death, that does not

appear to be the case; in the second, even if
it were, the fact of the verification of the
prognostic remains unaffected; besides, which
those who insist upon taking refuge in this
hypothesis, are bound to admit, that before
people living in the world, like Lord and
Lady A., could attach so much importance to
the prognostic as to entail such fatal effects,
they must have had very good reason for
believing in it.

The legend connected with the drummer is,
that either himself, or some officer whose
emissary he was, had become an object of
jealousy to a former Lord A., and that he was
put to death by being thrust into his own drum
and flung from the window of the tower in
which Miss D's room was situated. It is
said, that he threatened to haunt them if they
took his life; and he seems to have been as
good as his word, having been heard several
times in the memory of persons yet living.

There is a curious legend attached to the
family of G. of R., to the effect, that when a
lady is confined in that house, a little old
woman enters the room when the nurse is
absent and strokes down the bed-clothes; after
which, the patient according to the technical
phrase " never does any good," and dies.

Whether the old lady has paid her visits or not, I do not know; but it is remarkable, that the results attending several late confinements there, have been fatal.

There was a legend in a certain family, that a single swan was always seen on a particular lake before a death. A member of this family told me, that on one occasion, the father, being a widower, was about to enter into a second marriage. On the wedding day, his son, appeared so exceedingly distressed, that the bridegroom was offended, and expostulating with him, was told by the young man, that his low spirits were caused by his having seen the swan. He, the son, died that night quite unexpectedly.

Besides Lord Littleton's dove, there are a great many very curious stories recorded in which birds have been seen in a room when a death was impending ; but the most extraordinary prognostic I know is that of the black dog, which seems to be attached to some families :—

A young lady of the name of P. not long since was sitting at work, well and cheerful, when she saw to her great surprise a large black dog close to her. As both door and window were closed, she could not understand

how he had got in, but when she started up
to put him out, she could no longer see him.
Quite puzzled and thinking it must be some
strange illusion, she set down again and went
on with her work, when, presently, he was
there again.   Much alarmed she now ran
out and told her mother, who said she must
have fancied it, or be ill. She declared neither
was the case, and to oblige her, the mother
agreed to wait outside the door, and if she
saw it again, she was to call her. Miss P.
re-entered the room and presently there was
the dog again; but when she called her mother,
he disappeared.   Immediately afterwards, the
mother was taken suddenly ill and died.
Before she expired, she said to her daughter,
"Remember the black dog!"

I confess, I should have been much disposed to think this a spectral illusion, were
it not for the number of corroborative instances;
and I have only this morning read in the
review of a work called "The Unseen World,"
just published, that there is a family in Cornwall who are also warned of an approaching
death by the apparition of a black dog, and a
very curious example is quoted, in which a
lady newly married into the family, and
knowing nothing of the tradition, came down

from the nursery to request her husband would go up and drive away a black dog that was lying on the child's bed. He went up and found the child dead.

I wonder if this phenomenon is the origin of the French phrase *bête noire*, to express an annoyance, or an augury of evil.

Most persons will remember the story of Lady Fanshawe, as related by herself; namely that whilst paying a visit to Lady Honor O'Brien, she was awakened the first night she slept there, by a voice, and on drawing back the curtain she saw a female figure standing in the recess of the window attired in white, with red hair and a pale and ghastly aspect; " She looked out of the window," says Lady Fanshawe, " and cried in a loud voice, such as I never before heard, ' A Horse! A Horse! A Horse!' and then with a sigh, which rather resembled the wind than the voice of a human being, she disappeared. Her body appeared to me rather like a thick cloud than a real solid substance. I was so frightened," she continues, " that my hair stood on end and my night cap fell off. I pushed and shook my husband, who had slept all the time, and who was very much surprised to find me in such a fright, and still more so when I told him the

cause of it and showed him the open window.
Neither of us slept any more that night, but
he talked to me about it, and told me how
much more frequent such apparitions were in
that country than in England." This was,
however, what is called a Banshee, for in the
morning Lady Honor came to them, to say,
that one of the family had died in the night,
expressing a hope that they had not been
disturbed; for said she, whenever any of the
O'Briens is on his death-bed it is usual for a
woman to appear at one of the windows every
night till he expires; but when I put you into
this room I did not think of it." This apparition was connected with some sad tale of
seduction and murder.

I could relate many more instances of this
kind, but I wish as much as possible to avoid
repeating cases already in print; so I will conclude this chapter with the following account
of Pearlin Jean, whose persevering annoyances,
at Allanbank were so thoroughly believed and
established, as to have formed at various times
a considerable impediment to letting the place.
I am indebted to Mr. Charles Kirkpatrick
Sharpe for the account of Jean and the anecdote that follows.

A housekeeper, called Betty Norrie, that

lived many years at Allanbank, declared she
and various other people had frequently seen
Jean, adding, that they were so used to her,
as to be no longer alarmed at her noises.

" In my youth," says Mr. Sharpe, " Pearlin
Jean was the most remarkable ghost in Scotland, and my terror when a child. Our old nurse,
Jenny Blackadder, had been a servant at Allanbank, and often heard her rustling in silks up
and down stairs, and along the passage. She
never saw her; but her husband did.

" She was a French woman, whom the first
baronet of Allanbank, then Mr. Stuart, met
with at Paris, during his tour to finish his
education as a gentleman. Some people said
she was a nun; in which case she must have
been a Sister of Charity, as she appears not to
have been confined to a cloister. After some
time, young Stuart either became faithless to
the lady, or was suddenly recalled to Scotland by his parents, and had got into his carriage, at the door of the hotel, when his Dido
unexpectedly made her appearance, and step-
ping on the fore-wheel of the coach to address
her lover, he ordered the postilion to drive on;
the consequence of which was that the lady
fell, and one of the wheels going over her forehead, killed her.

"In a dusky autumnal evening, when Mr. Stuart drove under the arched gateway of Allanbank, he perceived Pearlin Jean sitting on the top, her head and shoulders covered with blood.

"After this, for many years, the house was haunted; doors shut and opened with great noise at midnight; the rustling of silks, and pattering of high-heeled shoes were heard in bed-rooms and passages. Nurse Jenny said there were seven ministers called in together at one time, to *lay* the spirit; 'but they did no mickle good, my dear.'

"The picture of the ghost was hung between those of her lover and his lady, and kept her comparatively quiet; but when taken away, she became worse natured than ever. This portrait was in the present Sir J. G.'s possession. I am unwilling to record its fate.

"The ghost was designated Pearlin, from always wearing a great quantity of that sort of lace.*

"Nurse Jenny told me that when Thomas Blackadder was her lover (I remember Thomas very well), they made an assignation to meet one moonlight night in the orchard at Allan-

* A species of lace made of thread. *Jamieson.*

bank. True Thomas, of course, was the first comer; and seeing a female figure, in a light coloured dress, at some distance, he ran for ward with open arms to embrace his Jenny; lo, and behold! as he neared the spot where the figure stood, it vanished; and presently he saw it again at the very end of the orchard, a considerable way off. Thomas went home in a fright; but Jenny, who came last, and saw nothing, forgave him, and they were married.

" Many years after this, about the year 1790, two ladies paid a visit at Allanbank—I think the house was then let—and passed a night there. They had never heard a word about the ghost; but they were disturbed the whole night with something walking backwards and forwards in their bed-chamber. This I had from the best authority.

" Sir Robert Stuart was created a baronet in the year 1687.

" Lady Stapleton, grandmother of the late Lord Le Despencer, told me, that the night Lady Susan Fane, Lord Westmoreland's daughter, died in London, she appeared to her father, then at Merriworth, in Kent. He was in bed, but had not fallen asleep. There was a light in the room; she came in, and sat down on a chair at the foot of the bed. He

said to her, 'Good God, Susan! how came you
here? What has brought you from town?'
She made no answer; but rose directly, and
went to the door, and looked back towards him
very earnestly; then she retired, shutting the
door behind her. The next morning he had
notice of her death. This, Lord Westmoreland himself told to Lady Stapleton, who was
by birth a Fane, and his near relation."

CHAPTER V.

APPARITIONS SEEKING THE PRAYERS OF THE
LIVING.

WITH regard to the appearance of ghosts, the
frequency of haunted houses, presentiments,
prognostics, and dreams, if we come to enquire
closely, it appears to me that all parts of the
world are much on an equality, only that where
people are most engaged in business or pleasure
these things are, in the first place, less thought
of, and less believed in, consequently less observed; and when they *are* observed, they are
readily explained away: and, in the second
place, where the external life—the life of the
brain, wholly prevails, either they do not

happen, or they are not perceived—the rapport not existing, or the receptive faculty being obscured.

But although the above phenomena seem to be equally well known in all countries, there is one peculiar class of apparitions, of which I meet with no records but in Germany. I allude to ghosts, who, like those described in the "Seeress of Prevorst," seek the prayers of the living. In spite of the positive assertions of Kerner, Eschenmayer, and others, that after neglecting no means to investigate the affair, they had been forced into the conviction, that the spectres that frequented Frederica Hauffe were not subjective illusions, but real outstanding forms, still, as she was in the somnambulic state, many persons remain persuaded that the whole thing was delusion. It is true, that as those parties were not there, and as all those who did go to the spot, came to a different conclusion, this opinion being only the result of preconceived notions or prejudices, and not of calm investigation, is of no value whatever; nevertheless, it is not to be denied, that these narrations are very extraordinary; but perplexing as they are, they by no means stand alone. I find many similar ones noticed in various works, where there has been no somnambule in ques-

tion. In all cases, these unfortunate spirits
appear to have been waiting for some one with
whom they could establish a rapport, so as
to be able to communicate with them; and
this waiting has sometimes endured a century
or more. Sometimes, they are seen by only
one person, at other times, by several, with
varying degrees of distinctness, appearing to
one as a light, to another as a shadowy figure,
and to a third as a defined human form. Other
testimonies of their presence, as sounds, footsteps, lights, visible removing of solid articles
without a visible agent, odours, &c. are generally perceived by many; in short, the sounds
seem audible to all who come to the spot, with
the exception of the voice, which, in most instances, is only heard by the person with whom
the rapport is chiefly established. Some cases
are related, where a mark like burning is left
on the articles seen to be lifted. This is an old
persuasion, and has given rise to many a joke;
but upon the hypothesis I have offered, the
thing is simple enough; the mark will probably
be of the same nature as that left by the electrical fluid; and it is this particular, and the
lights that often accompany spirits, that have
caused the notion of material flames, sulphur,
brimstone, and so forth, to be connected with

the idea of a future state. According to our views, there can be no difficulty in conceiving, that a happy and blessed spirit would emit a mild radiance; whilst anger or malignity would necessarily alter the character of the effulgence.

As whoever wishes to see a number of these cases may have recourse to my translation of the "Seeress of Prevorst," I will here only relate one, of a very remarkable nature, that occurred in the prison of Weinsberg, in the year 1835.

Dr. Kerner, who has published a little volume, containing a report of the circumstances, describes the place where the thing happened, to be such an one as negatives, at once, all possibility of trick or imposture. It was in a sort of block-house or fortress---a prison within a prison—with no windows but what looked into a narrow court or passage, which passage was closed with several doors. It was on the second floor; the windows being high up, heavily barred with iron, and immoveable without considerable mechanical force. The external prison is surrounded by a high wall and the gates are kept closed day and night. The prisoners in different apartments are of course, never allowed to communicate with each other, and the deputy-governor of the prison and his family, consisting of a wife and niece, and one maid-servant, are described as people of unimpeachable respectability and veracity. As depositions regarding this affair were laid before the magistrates; it is on them I found my narration.

On the 12th September, 1835, the deputy-governor, or keeper, of the jail, named Mayer, sent in a report to the magistrates, that a woman called Elizabeth Eslinger, was every night visited by a ghost, which generally came about eleven o'clock, and which left her no rest, as it said, she was destined to release it; and it always invited her to follow it; and as she would not, it pressed heavily on her neck and side, till it gave her pain. The persons confined with her, pretended also to have seen this apparition.

(Signed)   " MAYER."

A woman named Rosina Schahl, condemned to eight days' confinement for abusive language, deposed, that about eleven o'clock, Eslinger began to breathe hard, as if she was suffocating; she said, a ghost was with her, seeking his salvation. "I did not trouble myself about it, but told her to wake me when it came again.   Last night I saw a shadowy form, between four and five feet high, standing near the bed; I did not see it move. Eslinger breathed very hard, and complained of a pressure on the side. For several days she has neither ate or drank any thing.

(Signed)   "SCHAHL."

" COURT RESOLVES

" That Eslinger is to be visited by the prison physician, and a report made as to her mental and bodily health.

" Signed by the Magistrates,

" ECKHARDT.
" THEURER.
" KNORR.

"REPORT.

" Having examined the prisoner, Elizabeth Eslinger, confined here since the beginning of September, I found her of sound mind, but possessed with one fixed idea, namely, that she is, and has been for a considerable time, troubled by an apparition, which leaves her no rest, coming chiefly by night, and requiring her prayers to release it. It visited her before she came to the prison, and was the cause of the offence that brought her here. Having now, in compliance with the orders of the Supreme Court, observed this woman for

eleven weeks, I am led to the conclusion that
there is no deception in this case, and also that
the persecution is not a mere monomaniacal idea
of her own, and the testimony not only of her
fellow prisoners, but that of the deputy-
governor's family, and even of persons in
distant houses, confirms me in this persuasion.

"Eslinger is a widow, aged thirty-eight
years, and declares that she never had any
sickness whatever; neither is she aware of any
at present; but she has always been a ghost-
seer, though never till lately had any communication with them. That, now, for eleven
weeks that she has been in the prison, she is
nightly disturbed by an apparition, that had
previously visited her in her own house, and
which had been once seen, also, by a girl of
fourteen, a statement which this girl confirms.
When at home, the apparition did not appear
in a defined human form, but as a pillar of cloud,
out of which proceeded a hollow voice, signifying
to her that she was to release it, by her prayers
from the cellar of a woman in Wimmenthal,
named Singhaasin, whither it was banished,
or whence it could not free itself. She, Eslinger, says that she did not then venture to
speak to it, not knowing whether to address it
as *Sie, Ihr,* or *Du,* (that is, whether she

should address it in the second or third person, which custom, amongst the Germans, has rendered a very important point of etiquette. It is to be remembered that this woman was a peasant, without education, who had been brought into trouble by taking to treasure-seeking, a pursuit in which she hoped to be assisted by this spirit. This digging for buried treasure is a strong passion in Germany).

" The ghost now comes in a perfect human shape, and is dressed in a loose robe, with a girdle, and has on its head a four-cornered cap. It has a projecting chin and forehead, fiery, deep-set eyes, a long beard, and high cheek bones, which look as if they were covered with parchment. A light radiates about and above his head, and in the midst of this light she sees the outlines of the spectre.

" Both she and her fellow-prisoners declare, that this apparition comes several times in a night, but always between the evening and morning bell. He often comes through the closed door or window, but they can then see neither door or window, or iron bars, they often hear the closing of the door, and can see into the passage when he comes in or out that way, so that if a piece of wood lies there, they see it. They hear a shuffling in the passage as

he comes and goes. He most frequently
enters by the window, and they then hear a
peculiar sound there. He comes in quite
erect. Although their cell is entirely closed,
they feel a cool wind* when he is near them.
All sorts of noises are heard, particularly a
crackling. When he is angry, or in great
trouble, they perceive a strange mouldering,
earthy smell. He often pulls away the coverlit,
and sits on the edge of the bed. At first the
touch of his hand was icy cold, since he
became brighter, it is warmer; she first saw
the brightness at his finger-ends, it afterwards spread further. If she stretches out her
hand she cannot feel him, but when he touches
her, she feels it; he sometimes takes her hands
and lays them together, to make her pray.
His sighs and groans are like a person in
despair; they are heard by others as well as
Eslinger. Whilst he is making these sounds,
she is often praying aloud, or talking to her
companions, so they are sure it is not she who
makes them. She does not see his mouth
move when he speaks. The voice is hollow
and gasping. He comes to her for prayers, and
he seems to her like one in a mortal sickness,
who seeks comfort in the prayers of others.

* It is to be observed that this is the sensation asserted to be
felt by Reichenbach's patients on the approach of the magnet.

He says he was a Catholic priest in Wimmenthal, and lived in the year 1414."

(Wimmenthal is still Catholic; the woman Eslinger herself is a Lutheran, and belongs to Backnang.)

" He says, that amongst other crimes, a fraud committed conjointly with his father, on his brothers, presses sorely on him; he cannot get quit of it; it obstructs him. He always entreated her to go with him to Wimmenthal, whither he was banished, or consigned, and to pray there for him.

" She says, she cannot tell whether what he says is true; and does not deny that she thought to find treasures by his aid. She has often told him that the prayers of a sinner, like herself, cannot help him, and that he should seek the Redeemer; but he will not forbear his entreaties. When she says these things, he is sad, and presses nearer to her, and lays his head so close that she is obliged to pray into his mouth. *He seems hungry for prayers.* She has often felt his tears on her cheek and neck; they felt icy cold; but the spot soon after burns, and they have a bluish red mark. (These marks are visible on her skin.)

" One night this apparition brought with him a large dog, which leapt on the beds,

and was seen by her fellow prisoners also, who
were much terrified, and screamed. The ghost
however spoke, and said, ' Fear not ; this is my
father.' He had since brought the dog with
him again, which alarmed them dreadfully,
and made them quite ill.

"Both Mayer and the prisoners asserted,
that Eslinger was scarcely seen to sleep, either
by night or day, for ten weeks, she ate very
little, prayed continually, and appeared very
much wasted and exhausted. She said that
she saw the spectre alike, whether her eyes
were open or closed, which showed that it was
a magnetic perception, and not *seeing* by her
bodily organs. It is remarkable that a cat belonging to the jail, being shut up in this
room, was so frightened when the apparition
came, that it tried to make its escape by flying
against the walls ; and finding this impossible,
it crept under the coverlet of the bed, in extreme terror. The experiment was made again,
with the same result ; and after this second
time, the animal refused all nourishment,
wasted away, and died.

"In order to satisfy myself," says Dr.
Kerner, "of the truth of these depositions, I
went to the prison on the night of the 15th
October, and shut myself up without light in

Eslinger's cell. About half-past eleven I heard a sound as of some hard body being flung down; but not on the side where the woman was, but the opposite; she immediately began to breathe hard, and told me the spectre was there. I laid my hand on her head, and adjured it as an evil spirit to depart. I had scarcely spoken the words when there was a strange rattling, crackling noise all round the walls, which finally seemed to go out through the window; and the woman said that the spectre had departed.

"On the following night it told her, that it was grieved at being addressed as an evil spirit, which it was not; but one that deserved pity; and that what it wanted, was prayers and redemption.

"On the 18th October, I went to the cell again, between ten and eleven, taking with me my wife, and the wife of the keeper, Madame Mayer. When the woman's breathing showed me the spectre was there, I laid my hand on her, and adjured it, in gentle terms, not to trouble her further. The same sort of sound as before commenced, but it was softer, and this time continued all along the passage, where there was certainly nobody. We all heard it.

"On the night of the 20th, I went again,

with Justice Heyd.   We both heard sounds when the spectre came, and the woman could not conceive why we did not see it.   We could not; but we distinctly felt a cool wind blowing upon us, when, according to her account it was near, although there was no aperture by which air could enter."

On each of these occasions, Dr. K. seems to have remained about a couple of hours.

Madame Mayer now resolved to pass a night in the cell, for the purpose of observation; and she took her niece, a girl aged nineteen, with her; her report is as follows:

"It was a rainy night, and in the prison pitch dark; my niece slept sometimes; I remained awake all night, and mostly sitting up in bed.

"About midnight, I saw a light come in at the window; it was a yellowish light and moved slowly; and though we were closely shut in, I felt a cool wind blowing on me.   I said to the woman, 'The ghost is here, is he not?'   She said, 'Yes,' and continued to pray, as she had been doing before.   The cool wind and the light now approached me; my coverlet was quite light, and I could see my hands and arms, and at the same time I perceived an indescribable odour of putrefaction, my face felt as if ants were running over it.   (Most of the prisoners described themselves as feeling the same sensation when the spectre was there.) Then the light moved about, and went up and down the room; and on the door of the cell, I saw a number of little glimmering stars, such as I had never before seen. Presently, I and my niece heard a voice which I can compare to nothing I ever heard before. It was not like a human voice. The words and sighs sounded as if they were drawn up out of a deep hollow, and appeared to ascend from the floor to the roof in a column; whilst this voice spoke, the woman was praying aloud: so I was sure it did not proceed from her. No one could produce such a sound. They were strange superhuman sighs, and entreaties for prayers and redemption.

" It is very extraordinary, that whenever the ghost spoke, I always *felt it beforehand.* (Proving that the spirit had been able to establish a rapport with this person. She was in a magnetic relation to him.) We heard a crackling in the room also; I was perfectly awake, and in possession of my senses, and we are ready to make oath to having seen and heard these things."

On the 9th of December Madame Mayer spent the night again in the cell, with her

niece and her maid-servant; and her report is
as follows:

"It was moonlight, and I sat up in bed all
night, watching Eslinger. Suddenly I saw a
white shadowy form, like a small animal, cross
the room. I asked her what it was; and she
answered, 'Don't you see its a lamb? It often
comes with the apparition.' We then saw a
stool, that was near us, lifted and set down again
on its legs. She was in bed, and praying the
whole time. Presently, there was such a noise
at the window that I thought all the panes
were broken. She told us it was the ghost,
and that he was sitting on the stool. We then
heard a walking and shuffling up and down,
although I could not see him; but presently, I
felt a cool wind blowing on me, and out of
this wind the same hollow voice I had heard
before, said, 'In the name of Jesus, look on me!'

"Before this, the moon was gone, and it was
quite dark; but when the voice spoke to me, I
saw a light around us, though still no form.
Then there was a sound of walking towards
the opposite window, and I heard the voice
say, 'Do you see me now?' And then for
the first time I saw a shadowy form, stretching
up as if to make itself visible to us, but could
distinguish no features.

"During the rest of the night, I saw it repeatedly, sometimes sitting on the stool, and at others moving about; and I am perfectly certain that there was no moon-light now, nor any other light from without. How I saw it, I cannot tell; it is a thing not to be described.

"Eslinger prayed the whole time, and the more earnestly she did so, the closer the spectre went to her. It sometimes sat upon her bed.

"About five o'clock, when he came near to me, and I felt the cool air, I said, 'Go to my husband, in his chamber, and leave a sign that you have been there!' He answered distinctly, 'Yes.' Then we heard the door, which was fast locked, open and shut; and we saw the shadow float out (for he floated rather than walked), and we heard the shuffling along the passage.

"In a quarter of an hour we saw him return, entering by the window; and I asked him if he had been with my husband, and what he had done. He answered by a sound like a short, low, hollow laugh. Then he hovered about without any noise, and we heard him speaking to Eslinger, whilst she still prayed aloud. Still, as before, I always knew when he was going to speak. After six o'clock, we saw him no more. In the morning, my husband mentioned with great surprise that his chamber door, which he was sure he had fast bolted and locked, even taking out the key when he went to bed, he had found wide open."

On the 24th, Madame Mayer passed the night there again, but on this occasion she only saw a white shadow coming and going, and standing by the woman, who prayed unceasingly. She also heard the shuffling.

Between prisoners and the persons in authority who went to observe, the number of those who testify to this phenomenon is considerable; and, although the amount of what was perceived varied according to the receptivity of the subject in each case, the evidence of all is perfectly coincident as to the character of the phenomena. Some saw only the light; others distinguished the form in the midst of it; all heard the sound, and perceived the mouldering earthy smell.

That the receptivity of the women was greater than that of the men, after what I have elsewhere said, should excite no surprise; the preponderance of the sympathetic system in them being sufficient to account for the difference.

Frederica Follen, from Lowenstein, who was eight weeks in the same cell with

Eslinger, was witness to all the phenomena, though she only once arrived at seeing the spectre in its perfect human form, as the latter saw it. But it frequently spoke to her, bidding her amend her life; and remember that it was one who had tasted of death that give her this counsel. This circumstance had a great effect upon her.

When any of them swore, the apparition always evinced much displeasure, grasped them by the throat, and forced them to pray. Frequently, when he came or went, they said it sounded like a flight of pigeons.

Catherine Sinn, from Mayenfels, was confined in an adjoining room for a fortnight. After her release, she was interrogated by the minister of her parish, and deposed that she had known nothing of Eslinger, or the spectre, " but every night, being quite alone, I heard a rustling and a noise at the window, which looked only into the passage. I felt and heard, though I could not see anybody, that some one was moving about the room; these sounds were accompanied by a cool wind, though the place was closely shut up. I heard also a crackling, and a shuffling, and a sound as if gravel were thrown; but could find none in the morning. Once it seemed to me that a

hand was laid softly on my forehead. I did
not like staying alone, on account of these
things; and begged to be put into a room with
others; so I was placed with Eslinger and
Follen. The same things continued here, and
they told me about the ghost; but not being
alone, I was not so frightened. I often heard
him speak; it was hollow and slow; not like
a human voice; but I could seldom catch the
words; when he left the prison, which was
generally about five in the morning, he used to
say, 'Pray!' and when he did so, he would
add, 'God reward you!' I never saw him distinctly till the last morning I was there; then,
I saw a white shadow standing by Eslinger's
bed.   (Signed)   CATHERINE SINN,

"Minister Binder, Mayenfels."

It would be tedious were I to copy the depositions of all the prisoners; the experience
of most of them being similar to the above.
I will therefore content myself with giving an
abstract of the most remarkable particulars.

Besides the crackling, rustling as of paper,
walking, shuffling, concussions of the windows,
and of their beds, &c. &c. they heard sometimes a fearful cry; and not unfrequently the
bed-coverings were pulled from them; it

appearing to be the object of the spirit to manifest himself thus to those to whom he could not make himself visible; and as I find this pulling off the bed-clothes and heaving up the bed, as if some one were under it, repeated in a variety of cases, foreign and English, I conclude the motive to be the same. Several of the women heard him speak.

All these depositions are contained in Dr. Kerner's report to the magistrates; and he concludes by saying, that there can be no doubt of the fact of the woman Elizabeth Eslinger suffering these annoyances, by whatever name people may choose to call them.

Amongst the most remarkable phenomena, is the real or apparent opening of the door ; so that they could see what was in the passage. Eslinger said that the spirit was often surrounded by a light, and his eyes looked fiery, and there sometimes came with him two lambs, which occasionally appeared as stars. He often took hold of Eslinger, and made her sit up, put her hands together, that she might pray ; and once he appeared to take a pen and paper from under his gown, and wrote, laying it on her coverlet.

It is extremely curious, that on two occasions Eslinger saw Dr. Kerner and Justice

Heyd enter with the ghost, when they were
not there in the body, and both times Heyd
was enveloped in a black cloud. The ghost,
on being asked, told Eslinger that the cloud
indicated that trouble was impending. A few
days afterwards, his child died very unexpectedly, and Dr. Kerner now remembered, that
the first time Eslinger said she had seen Heyd
in this way, his father had died directly afterwards. Kerner attended both patients, and
was thus associated in the symbol. Follen
also saw these two images, and spoke, believing the one to be Dr. Kerner himself.

On other occasions, she saw strangers come
in with the ghost, whom afterwards, when
they *really* came in the body, she recognised;
this seems to have been a sort of second sight.

Dr. K. says, I think justly enough, that if
Eslinger had been feigning, she never would
have ventured on what seemed so improbable.

Some of the women, after the spectre had
visibly leant over them, or had spoken into
their ears, were so affected by the odour he
diffused, that they vomited, and could not eat
till they had taken an emetic, and those parts
of their persons that he touched became painful
and swollen, an effect I find produced in
numerous other instances.

The following particulars are worth observing, in the evidence of a girl sixteen years
of age, called Margaret Laibesberg, who was
confined for ten days for plucking some grapes
in a vineyard. She says, she knew nothing
about the spectre, but that she was greatly
alarmed, the first night, at hearing the door
burst open, and something come shuffling in.
Eslinger bade her not fear, and said that it
would not injure her. The girl, however,
being greatly terrified every night, and hiding
her head under the bed-clothes, on the fourth
Eslinger got out of her own bed, and coming
to her, said, " Do, in the name of God, look at
him ! He will do you no harm, I assure you."
" Then," says the girl, " I looked out from
under the clothes, and I saw two white forms,
like two lambs—so beautiful that I could have
looked at them for ever. Between them
stood a white, shadowy form, as tall as a man,
but I was not able to look longer, for my eyes
failed me." The terrors of this girl were so
great, that Eslinger had repeatedly occasion
to get out of bed and fetch her to lie with
herself. When she could be induced to look,
she always saw the figure, and he bade her
also pray for him. Whenever he touched her,
which he did on the forehead and eyes, she

felt pain, but says nothing of any subsequent
swelling. Both this girl and another, called
Neidhardt, who was brought in on the last
day of Margaret L.'s imprisonment, testified,
that, on the previous night, they had heard
Eslinger ask the ghost, " Why he looked so
angry ?" And that they had heard him
answer, that it was " Because she had, on the
preceding night, neglected to pray for him as
much as usual," which neglect arose from two
gentlemen having passed the night in the cell.

When on the tenth day the girl Margaret L.
was released, she said that there was something so awful to her in this apparition, that
she had firmly resolved and vowed to be
pious, and lead, henceforth, a virtuous life.

Some of them seem to have felt little alarm ;
Maria Bar, aged forty-one, said, " I was not
afraid, for I have a good conscience." The
offences for which these women were confined
appear to have been very slight ones, such as
quarrelling, and so forth.

In a room that opened into the same passage, men were shut up for disputing with the
police, neglect of regulations, and similar mis-
demeanours. These persons not only heard
the noises as above described, such as the
walking, shuffling, opening and shutting

the door, &c. &c., but some of them saw
the ghost. Christian Bauer deposed, that he
had never heard anything about the ghost, but
that, being disturbed by a knocking and rustling towards three o'clock on the second
morning of his incarceration, he looked up,
and saw a white figure bending over him, and
heard a strange hollow voice say, " You must
needs have patience !" He said, he thought
it must be his grandfather, at which Stricker,
his companion, laughed. Stricker deposed,
that he heard a hollow voice say, " You must
needs have patience," and that Bauer told him
that there was a white apparition near him,
and that he supposed it was his grandfather.
Bauer said, that he was frightened the first
night, but got used to it, and did not mind

It is worthy of observation, that when they,
heard the door of the women's room open, they
also heard the voice of Eslinger praying, which
seems as if the door not only appeared to open,
but actually did so. We have already seen
that this spirit could open doors. In the
" Seeress of Prevorst," the doors were constantly *visibly* and *audibly* opened, as by an
unseen hand, when she saw a spectre enter ;
and I know to an absolute certainty, that the
same phenomenon takes place in a house not

far from where I am writing; and this, sometimes, when there are two people sleeping in
the room—a lady and gentleman. The door
having been fast locked when they went to
bed, the room thoroughly examined, and every
possible precaution taken, for they are unwilling to believe in the spiritual character of
the disturbances that annoy them—they are
aroused by a consciousness that it is opening,
and they do find it wide open, on rising to
investigate the fact.

One of the most remarkable proofs, either of
the force of volition or of the electrical powers
of the apparition that haunted Eslinger, or
else of his power to imitate sounds, was the
real, or apparent, violent shaking of the heavy,
iron-barred window, which it is asserted the
united efforts of six men could not shake at all,
when they made the experiment.

The Supreme Court having satisfied itself
that there was no imposture in this case, it
was proposed that some men of science should
be invited to investigate the strange phenomenon, and endeavour, if possible, to explain
it. Accordingly, not only Dr. Kerner himself and his son, but many others, passed
nights in the prison, for this purpose.
Amongst these, besides some ministers of the

Lutheran Church, there was an engraver, called Duttenhofer; Wagner, an artist; Kapff, professor of mathematics at Heilbronn; Fraas, a barrister; Doctors Seyffer and Sicherer, physicians; Heyd, a magistrate; Baron von Hugel, &c. &c.; but their perquisitions elicited no more than has been already narrated; all heard the noises, most of them saw the lights, and some saw the figure. Duttenhofer and Kapff saw it without a defined outline; it was itself bright, but did not illuminate the room. Some of the sounds appeared to them like the discharging of a Leyden jar. There was also a throwing of gravel, and a heavy dropping of water, but neither to be found. Professor Klapff says, that he was quite cool and self-possessed, till there was such a violent concussion of the heavy barred window, that he thought it must have come in; then both he and Duttenhofer felt horror-struck.

As they could not see the light emitted by the spectre when the room was otherwise lighted, they were in the dark; but they took every care to ascertain that Eslinger was in her bed whilst these things were going on. She prayed aloud the whole time, unless when speaking to them.

By the morning, she used to be dreadfully exhausted, from this continual exertion.

It is also mentioned, that the straw on which she lay was frequently changed and examined, and every means taken to ascertain that there was nothing whatever in her possession that could enable her to perform any sort of jugglery. Her fellow prisoners were also invited to tell all they knew or could discover; and a remission of their sentences promised to those who would make known the imposition, if there were one.

Dr. Sicherer, who was accompanied by Mr. Frass, says, that having heard of these phenomena, which he thought the more unaccountable from the circumstances of the woman's age and condition, &c., she being a healthy hard-working person, aged thirty-eight, who had never known sickness, he was very desirous of enquiring personally into the affair.

Whilst they were in the court of the prison, waiting for admittance, they heard extraordinary noises, which could not be accounted for, and during the night there was a repetition of those above described ; especially the apparent throwing of gravel, or peas, which seemed to fall so near him that he involuntarily covered his face. Then followed the feeling of a cool wind; and then the oppressive odour, for which, he says, he can find no comparison, and which almost took away his breath.   He was perfectly satisfied that it was no smell originating in the locality or the state of the prison.   Simultaneously with the perception of this odour, he saw a thick grey cloud, of no defined shape, near Eslinger's bed.   When this cloud disappeared, the odour was no longer perceptible.   It was a fine moonlight night, and there was light enough in the room to distinguish the beds, &c.   The same phenomena recurred several times during the night; Eslinger was heard, each time the ghost was there, praying and reciting hymns.   They also heard her say, "Don't press my hands so hard together!" "Don't touch me!" and so forth. The voice of the spirit they did not hear.   Towards three or four o'clock, they heard heavy blows, footsteps, opening and shutting of the door, and a concussion of the whole house, that made them think it was going to fall on their heads.   About six o'clock, they saw the phantom again; and altogether these phenomena recurred at least ten times in the course of the night.

Dr. Sicherer concludes by saying, that he had undertaken the investigation with a mind entirely unprepossessed; and that in the report he made, at the desire of the Supreme Court, he had recorded his observations as conscientiously as if he had been upon a jury. He adds, that he had examined everything; and that neither in the person of the woman, nor in any other of the inmates of the prison, could he find the smallest grounds for suspicion, nor any clue to the mystery, which, in a scientific point of view appeared to him utterly inexplicable. Dr. Sicherer's report is dated Heilbronn, January 8, 1836.

Mr. Fraas, who accompanied him, confirms the above statement, in every particular; with the addition, that he several times saw a light of a varying circumference, moving about the room; and that it was whilst he saw this, that the woman told him the ghost was there. He also felt an oppression of the breath and a pressure on his forehead each time before the apparition came, especially once, when although he had carefully abstained from mentioning his sensations, she told him it was standing close at his head. He stretched out his hand; but perceived nothing, except a cool wind and an overpowering smell.

Dr. Seyffer being there, one night, with Dr. Kerner, in order to exclude the possibility of

light entering through the window, they
stopped it up. They, however, saw the phosphorescent light of the spectre, as before. It
moved quietly about; and remained a quarter
of an hour. The room was otherwise perfectly
dark; the sounds accompanying it were like
the dropping of water and the discharge of a
Leyden jar. They fully ascertained that these
phenomena did not proceed from the woman.

I have already given the depositions of
Madame Mayer, the wife of the deputy-
governor, or keeper of the prison, who is
spoken of as a highly respectable person.
Mayer himself, however, though quite unable
to account for all these extraordinary proceedings, found great difficulty in believing
that there was anything supernatural in the
affair; and he told Eslinger that if she wished
him to be convinced, she must send the ghost
to do it.

He says, " The night after I had said this, I
went to bed and to sleep, little expecting such
a visitor; but towards midnight, I was
awakened by something touching my left
elbow, this was followed by a pain; and in the
morning, when I looked at the place, I saw
several blue spots. I told Eslinger that this
was not enough, and that she must tell the

ghost to touch my other elbow. This was done on the following night, and, at the same time, I perceived a smell like putrefaction. The blue spots followed." (It will be remembered that Eslinger had blue spots also.)

Mayer continues to say, that the spectre made known its presence in his chamber by various sounds, such as were heard in the other part of the house. He never saw the figure distinctly, but his wife did; she always prayed when it was there. He, however, felt the cool wind that they all described.

The ghost told Eslinger that he should continue his visits to the prison after she had quitted it, and he did so. The second night after her release, they felt his approach, especially from the cool wind, and Madame Meyer desired him to testify his presence to her husband. Immediately there were sounds like a wind instrument, and these were repeated at her desire.

The prisoners also heard and felt the apparition after Eslinger's departure, and Mayer says he is perfectly assured, that in this jail, where the inmates were frequently changed, everybody was locked up, and every place thoroughly examined, it was utterly impossible for any trick to be played. Besides which, all parties agreed that the sounds were often of a description that could not have been produced by any known means.

But it was not to the prison alone that this apparition confined his visits. To whomsoever Eslinger sent him, he went, testifying his presence by the same signs as above described.

He visited the chambers of several of the magistrates, of a teacher called Neuffer, of the Referendary Burger, of a citizen named Rummel, and many others. Of these, some only perceived his presence by the noises, the cool air, the smell, or the touch; others saw the light also, and others perceived the figure with more or less distinctness.

A Mr. Dorr, of Heilbronn, seems to have scoffed very much at these rumours, and Dr. Kerner bade Eslinger ask the ghost to convince him, which she did.

Mr. Dorr says, " When I heard these things talked of, I always laughed at them, and was thought very sensible for so doing ; now I shall be laughed at in my turn, no doubt."

He then relates that, on the morning of the 30th December, 1835, he awoke, as usual, about five o'clock, and was thinking of some business he had in hand, when he became conscious that there was something near him, and he felt as if it blew cold upon him. He started up, thinking some animal had got into his room, but could find nothing. Next he heard a noise, like sparks from an electrical machine, and then a report close to his right ear. Had there been anything visible, it was light enough to see it. This report was frequently heard in the prison.

Wherever the apparition once made a visit, he generally continued to go for several successive nights. He also visited Professor Kapff, at Heilbronn, and Baron von Hugel at Eschenau, without being desired to do so by Eslinger; and Neuffer, whom he also went to, she knew nothing of.

When he visited Dr. Kerner's chamber, his wife, who had prided herself on her incredulity, and boasted of being born on St. Thomas's Day, was entirely converted, for she not only heard him, but saw him distinctly. He visited them for several nights, accompanied by the noises and the light.

One night, whilst lying awake observing these phenomena, they fancied they heard their horse come out of his stable, which was under their room. In the morning, he was found standing outside, with his halter on; it was not broken, and it was evident that the horse had not got loose by any violence. Moreover, the door of the stable was closed behind him, as it had been at night, when he was shut up.

Dr. K.'s sister, who came from a distance to visit them, had heard very little about this affair, yet she was awakened by a sound that seemed like some one trying to speak into her ear; and, looking up, she saw two stars, like those described by Margaret Laibesberg. She observed that they emitted no rays. She also felt the cool air, and perceived the corpse-like odour. This odour accompanied the ghost even when he appeared at Heilbronn.

It is remarkable that some of these persons, both men and women, felt themselves unable to move or call out, whilst the spectre was there; and that they were relieved the moment he went away. They appeared to be magnet- ised; but this feeling was by no means uni- versal. Many were perfectly composed and self-possessed the whole time, and made their observations to each other. All agreed that the speaking of the apparition seemed like that of a person making efforts to speak. Now, as we are to presume that he did not speak by means of organs, as we do, but that he imi- tated the sounds of words as he imitated other sounds, by some means with which we are unacquainted—for since the noises were heard by everybody within hearing, we must suppose that they actually existed—we, who know the extreme difficulty of imitating human speech, may conceive how this imitation should be very defective.

Dutthenhofer and others remarked, that there was no echo from the sounds, as well as that the phosphorescence shed no light around; and though the spectre could touch *them*, or produce the sensation that he did, they could not feel *him*; but, as in all similar cases, could thrust their hands through what appeared to be his body. The sensation of his falling tears, and the marks they left, seem most unaccountable; and yet, in the records of a ghost that haunted the Countess of Eberstein, in 1685, we find the same thing asserted. This account was made public by the authority of the Consistorial Court, and with the consent of the family.

At length, on the 11th of February, the ghost took his departure from Eslinger; at least, after that day he was no more seen or heard by her or anybody else. He had always entreated her to go to Wimmenthal, where he had formerly lived, to pray for him; and, after she was released from the jail, by the advice of her friends, she did it. Some of them accompanied her; and they saw the apparition near her whilst she was kneeling in the open air, though not all with equal distinctness. A very respectable woman, called Wörner, a stranger to Eslinger, whom she says she never saw or spoke to till that day, offered to make oath, that she had accompanied her to Wimmenthal, and that with the other friends, she had stood about thirty paces off, quite silent and still, whilst the woman knelt and prayed; and that she had seen the apparition of a man, accompanied by two smaller spectres, hovering near her. " When the prayer was ended, he went close to her, and there was a light like a falling star; then I saw something like a white cloud, that seemed to float away; and after that, we saw no more."

Eslinger had been very unwilling to undertake this expedition: she took leave of her children before she started, and evidently expected mischief would befall her; and now, on approaching her, they found her lying cold and insensible. When they had revived her, she told them, that on bidding her farewell, before he ascended, which he did, accompanied by two bright infantine forms, the ghost had asked her to give him her hand; and that after wrapping it in her handkerchief, she had complied. A small flame had arisen from the handkerchief when he touched it; and we found the marks of his fingers like burns, but without any smell. This, however, was not the cause of her fainting; but she had been terrified by a troop of frightful animals that she saw rush past her, when the spirit floated away.

From this time, nobody, either in the prison or out of it, was troubled with this apparition.

This is certainly a very extraordinary story; and what is more extraordinary, such cases do not seem to be very uncommon in Germany. I meet with many recorded: and an eminent German scholar, of my acquaintance, tells me that he has also heard of several, and was surprised that we have no similar instances here. If these things occurred merely amongst the Roman Catholics, we might be inclined to suppose they had some connexion with their notion of purgatory : but, on the contrary, it appears to be amongst the Lutheran population they chiefly occur; insomuch that it has even been suggested, that the omission of prayers for the dead, in the Lutheran Church, is the cause of the phenomenon. But, on the other hand, as in the present case, and in se-

veral others, the person that revisits the earth
was of the Catholic persuasion when alive, we
are bound to suppose that he had the benefit of
his own Church's prayers. I am here assuming that all the above strange phenomena
were really produced by the agency of an apparition : if they were not, what were they ? The
three physicians, who were amongst the
visitors, must have been perfectly aware of the
contagious nature of some forms of nervous
disorder, and from the previous incredulity of
two of them they must have been quite prepared to regard these phenomena from that
point of view ; yet they seem unable to bring
them under the category of sensuous illusions.

The apparently electrical nature of the lights,
and of several of the sounds, is very remarkable, as are also the swellings produced on some
of the persons by the touch of the ghost, which
remind us of Professor Hofer's case, mentioned
in a former chapter. The apparitions of the
dog and the lambs also, strange as they are,
are by no means isolated cases. These appearances seem to be symbolical : the father had
been evil, and had led his son to do evil, and
he appeared in the degraded form of a dog;
and the innocence of the children who had
been, probably, in some way wronged, was

symbolised by their appearing as lambs. "If
I had lived as a beast," said an apparition to
the Seeress of Prevorst, "I should appear
as a beast." These symbolical transfigurations
cannot appear very extravagent to those who
accept the belief of many theologians, that the
serpent of the garden of Eden was an evil spirit
incarnated in that degraded form.

How far the removal of the horse out of
his stable was connected with the rest of the
phenomena, it is impossible to say; but a
similar circumstance has very lately occurred
with regard to a dog that was locked up in the
house in this neighbourhood, which I have
several times alluded to, where footsteps and
rustlings are heard, doors are opened, and a
feeling that some one is blowing, or breathing
upon them, is felt by the inhabitants.

The holes burnt in the handkerchief are
also quite in accordance with many other relations of the kind, especially that of the
maid of Orlach, and also that of the Ham-
merschan family, mentioned in "Stilling's
Pneumatology," when a ghost who had been,
as he said, waiting one hundred and twenty
years for some one to release him by their
prayers, was seen to take a handkerchief, on
which he left the marks of his five fingers,
appearing like burnt spots. A Bible he

touched was marked in the same manner, and these two memento's of the apparition were carefully retained in the family. This particularity, also, reminds us of Lord Tyrone's leaving the marks of his hand on Lady Beresford's wrist, on which she ever afterwards wore a black ribbon. In several instances I find it reported that when an apparition is requested to render himself visible to, or to enter into communication with, other persons, besides those to whom he addresses himself, he answers that it is impossible; and in other cases that he could do it, but that the consequences to those persons would be pernicious. This, together with the circumstance of their waiting so long for the right person, tends strongly to support the hypothesis that an intense magnetic rapport is necessary to any facility of intercourse. It also appears that the power of establishing this rapport with one or more persons, varies exceedingly amongst these denizens of a spiritual world, some being only able to render themselves audible, others to render themselves visible to one person, whilst a few seem to possess considerably greater powers, or privileges.

Another particular to be observed is, that in many instances, if not in all, these spirits are what the Germans call *gebannt*, that is, *banned*, or *proscribed*, or, as it were, *tethered* to a certain spot, which they can occasionally leave, as Anton did the cellar at Wimmenthal, to which he was *gebannt*, but from which they cannot free themselves. To this spot they seem to be attached, as by an invisible chain, whether by the memory of a crime committed there, or by a buried treasure, or even by its being the receptacle of their own bodies. In short, it seems perfectly clear, admitting them to be apparitions of the dead, that, whatever the bond may be which keeps them down, they cannot quit the earth; they are, as St. Martin says, *remainers*, not *returners*, and this seems to be the explanation of haunted houses.

In the year 1827, Christian Eisengrun, a respectable citizen of Neckarsteinach, was visited by a ghost of the above kind, and the particulars were judicially recorded. He was at Eherbach, in Baden, working as a potter, which was his trade, in the manufactory of Mr. Gehrig, when he was one night awakened by a noise in his chamber, and, on looking up, he saw a faint light, which presently assumed a human form, attired in a loose gown; he could see no head. He had his own head under the clothes; but it presently spoke, and

told him that he was destined to release it, and
that for that purpose he must go to the
Catholic churchyard of Neckarsteinach, and
there, for twenty-one successive days, repeat
the following verse from the New Testament,
before the stone sepulchre there :—

"For what man knoweth the things of a
man, save the spirit of man which is in him?
So the things of God knoweth no man, but
the spirit of God."—1 Cor. ii. 11.

The ghost having repeated his visits and
his request, the man consulted his master
what he should do, and he advised him not
to trifle with the apparition, but to do what he
required, adding that he had known many
similar instances. Upon this, Eisengrun went
to Neckarsteinach, and addressed himself to
the Catholic priest there, named Seitz, who
gave him the same counsel, together with his
blessing, and also a hymn of Luther's, which
he bade him learn and repeat, as well as the
verse, when he visited the sepulchre.

As there was only one stone sepulchre in
the churchyard, Eisengrun had no difficulty in
finding it; and whilst he performed the service imposed on him by the ghost, the latter
stood on the grave with his hands folded, as if
in prayer; but when he repeated the hymn, he

moved rapidly backwards and forwards, but
still not overstepping the limits of the stone.
The man, though very frightened, persevered
in the thing for the time imposed, twenty-one
days; and during this period he saw the perfect
form of the apparition, which had no covering
on its head except very white hair. It always
kept its hands folded, and had large eyes, in
which he never perceived any motion; this
filled him with horror. Many persons went to
witness the ceremony.

The surviving nephews and nieces of the
apparition brought an action against Eisengrun, and they contrived to have him seized
and carried to the magistrate's house, one day,
at the time he should have gone to the churchyard. But the ghost came and beckoned, and
made signs to him to follow him, till the man
was so much affected and terrified, that he
burst into tears. The two magistrates could
not see the spectre, but feeling themselves
seized with a cold shudder, they consented to
his going.

He was then publicly examined in court,
together with the offended family and a number of witnesses, and the result was, that he
was permitted to continue the service for the
twenty-one days, after which he never saw or

heard more of the ghost, who had been formerly a rich timber-merchant.   The terror and
anxiety attendant on these daily visits to the
churchyard, affected Eisengrun so much, that
he was some time before he recovered his usual
health.   He had all his life been a ghost-seer,
but had never had communication with any
before this event.

The Catholic priest, in this instance, appears
to have been more liberal than the deceased
timber-merchant, for the latter did not seem to
like the Lutheran hymn, which the former
prescribed.   His dissatisfaction, however, may
have arisen from their making any addition to
the formula he had himself indicated.

CHAPTER VI.

THE POLTERGEIST OF THE GERMANS, AND POSSESSION.

WITH regard to the so-called *hauntings*, referred to in the preceding chapter, there seems reason to believe that the invisible guest was formerly a dweller upon earth, in the flesh, who is prevented by some circumstance which we are not qualified to explain, from pursuing the destiny of the human race, by entering freely into the next state prepared for him. He is like an unfortunate caterpillar that cannot entirely free itself from the integuments of its reptile life which chain it to the earth, whilst its fluttering wings vainly seek to bear it into the region to which it now belongs. But there is another kind of *haunting*, which is still more mysterious and strange, though by no means unfrequent, and which, from the odd, sportive mischievous nature of the disturbances created, one can scarcely reconcile to our notions of, what we understand by the term *ghost*. For in those cases where the unseen visitant appears to be the spirit of a person deceased, we see evidences of grief, remorse, and dissatisfaction, together with, in many instances, a disposition to repeat the acts of life, or, at least, to simulate a repetition of them: but there is nothing sportive or mischievous, nor, except where an injunction is disobeyed, or a request refused, are there generally any evidences of anger or malignity. But in the other cases alluded to, the annoyances appear rather like the tricks of a mischievous imp. I refer to what the Germans call the *Poltergeist*, or racketing spectre, for the phenomenon is known in all countries, and has been known in all ages.

Since hearing of the phenomenon of the electric girl, which attracted so much attention and occasioned so much controversy in Paris lately, and other similar cases, which have since, reached me, I feel doubtful whether some of these strange circumstances may not have been

connected with electricity in one form or
another. The famous story of what is familiarly called the Stockwell Ghost, for example, might possibly be brought under this
category. I have heard some people assert,
that the mystery of this affair was subsequently
explained away, and the whole found to be a
trick. But that is a mistake. Some years ago,
I was acquainted with persons whose parents
were living on the spot at the time, who knew
all the details, and to them it remained just as
great a mystery as ever. Not the smallest light
had ever been thrown upon it. People are so
glad to get rid of troublesome mysteries of this
description, that they are always ready to say,
" The trick has been found out !" and those
who pride themselves on not believing idle
stories, are to the last degree credulous when
" the idle story" flatters their scepticism.

The circumstances of the so-called Stockwell Ghost, which I extract from a report published at the time, are as follows.

The pamphlet was entitled—

" An Authentic, Candid, and Circumstantial
    Narrative, of the astonishing Transactions
    at Stockwell, in the County of Surrey, on
    Monday and Tuesday, the 6th and 7th days
    of January, 1772, containing a Series of the

most surprising and unaccountable Events
that ever happened, which continued from
first to last, upwards of Twenty Hours, and
at different places.

" Published with the consent and approbation
of the family and other parties concerned, to
authenticate which, the original copy is
signed by them.

"Before we enter upon a description of the
most extraordinary transactions that perhaps
ever happened, we shall begin with an account
of the parties who were principally concerned,
and in justice to them, give their characters,
by which means the impartial world may see
what credit is due to the following narrative.

" The events indeed are of so strange and
singular a nature, that we cannot be at all
surprised the public should be doubtful of the
truth of them, more especially as there have
been too many impositions of this sort; but,
let us consider. here are no sinister ends to be
answered, no contributions to be wished for,
nor would be accepted, as the parties are in
reputable situations and good circumstances,
particularly Mrs. Golding, who is a lady of an
independent fortune: Richard Fowler and his
wife might be looked upon as an exception to
this assertion ; but as their loss was trivial, they

must be left out of the question, except so far
as they appear corroborating evidences.

"Mr. Pain's maid lost nothing.

"How or by what means these transactions
were brought about has never transpired: we
have only to rest our confidence on the veracity
of the parties, whose descriptions have been
most strictly attended to, without the least de-
viation: nothing here offered is either exagge-
rated or diminished, the whole stated in the
clearest manner, just as they occurred : as such
only we lay them before the candid and impartial public.

"Mrs. Golding, an elderly lady, at Stockwell, in Surrey, at whose house the transactions
began, was born in the same parish (Lambeth)
has lived in it ever since, and has always been
well known, and respected as a gentlewoman
of unblemished honour and character. Mrs.
Pain, a niece of Mrs. Golding, has been married several years to Mr. Pain, a farmer, at
Brixton Causeway, a little above Mr. Angel's,
has several children, and is well known and
respected in the parish. Mary Martin, Mr.
Pain's servant, an elderly woman, has lived two
years with them, and four years with Mrs.
Golding, where she came from. Richard Fowler
lives almost opposite to Mr. Pain, at the Brick

Pound, an honest, industrious and sober man. And Sarah Fowler, wife to the above, is an industrious and sober woman.

" These are the subscribing evidences that we must rest the truth of the facts upon : yet there are numbers of other persons who were eye-witnesses of many of the transactions, during the time they happened, all of whom must acknowledge the truth of them.

" Another person who bore a principal part in these scenes was Ann Robinson, Mrs. Golding's maid, a young woman, about twenty years old, who had lived with her but one week and three days. So much for the *Historiæ Personæ*, and now for the narrative.

" On Monday, January the 6th, 1772, about ten o'clock in the forenoon, as Mrs. Golding was in her parlour, she heard the china and glasses in the back kitchen tumble down and break ; her maid came to her and told her the stone plates were falling from the shelf; Mrs. Golding went into the kitchen and saw them broke. Presently after, a row of plates from the next shelf fell down likewise, while she was there, and nobody near them ; this astonished her much, and while she was thinking about it, other things in different places began to tumble about, some of them breaking, at-

tended with violent noises all over the house;
a clock tumbled down and the case broke; a
lanthorn that hung on the stair-case was
thrown down and the glass broke to pieces;
an earthen pan of salted beef broke to pieces,
and the beef fell about: all this increased her
surprise, and brought several persons about
her, among whom was Mr. Rowlidge, a carpenter, who gave it as his opinion, that the
foundation was giving way and that the house
was tumbling down, occasioned by the too
great weight of an additional room erected
above: so ready are we to discover natural
causes for everything! But no such thing
happened as the reader will find, for whatever
was the cause, that cause ceased almost as
soon as Mrs. Golding and her maid left any
place, and followed them wherever they went.
Mrs. Golding ran into Mr. Gresham's house,
a gentleman living next door to her, where
she fainted.

"In the interim, Mr. Rowlidge and other
persons were removing Mrs. Golding's effects
from her house, for fear of the consequences
he had prognosticated. At this time all was
quiet; Mrs. Golding's maid remaining in her
house, was gone, up stairs, and when called
upon several times to come down, for fear of

the dangerous situation she was thought to be in, she answered very cooly, and after some time come down as deliberately, without any seeming fearful apprehensions.

"Mrs. Pain was sent for from Brixton Causeway, and desired to come directly, as her aunt was supposed to be dead ;—this was the message to her. When Mrs. Pain came, Mrs. Golding was come to herself, but very faint.

" Among the persons who were present, was Mr. Gardner, a surgeon, of Clapham ; whom Mrs. Pain desired to bleed her aunt, which he did ; Mrs. Pain asked him if the blood should be thrown away ; he desired it might not, as he would examine it when cold. These minute particulars would not be taken notice of, but as a chain to what follows. For the next circumstance is of a more astonishing nature than anything that had preceded it ; the blood that was just congealed, sprung out of the basin upon the floor, and presently after the basin broke to pieces : this china basin was the only thing broke belonging to Mr. Gresham : a bottle of rum that stood by it, broke at the same time.

" Amongst the things that were removed to Mr. Gresham's, was a tray full of china, &c. a japan bread basket, some mahogany waiters,

with some bottles of liquors, jars of pickles,
&c. and a pier glass, which was taken down
by Mr. Saville, (a neighbour of Mrs. Golding's);
he gave it to one Robert Hames, who laid it
on the grass-plat at Mr. Gresham's : but
before he could put it out of his hands, some
parts of the frame on each side flew off : it
rained at that time, Mrs. Golding desired it
might be brought into the parlour, where it
was put under a side-board, and a dressing-
glass along with it: it had not been there long
before the glasses and china which stood on
the side-board, began to tumble about and fall
down, and broke both the glasses to pieces.
Mr. Saville and others being asked to drink a
glass of wine or rum, both the bottles broke in
pieces before they were uncorked.

"Mrs. Golding's surprise and fear increasing,
she did not know what to do, or where to go;
wherever she and her maid were, these strange
destructive circumstances followed her, and
how to help or free herself from them, was not
in her power or any other person's present :
her mind was one confused chaos, lost to herself and everything about her, drove from her
own home, and afraid there would be none
other to receive her : at last she left Mr.
Gresham's, and went to Mr. Mayling's, a

gentleman at the next door, here she staid about three quarters of an hour, during which time nothing happened. Her maid staid at Mr. Gresham's, to put up what few things remained unbroke of her mistress's, in a back apartment, when a jar of pickles that stood upon a table turned upside down, then a jar of raspberry jam broke to pieces, next two mahogany waiters and a quadrille-box likewise broke in pieces.

" Mrs. Pain, not choosing her aunt should stay too long at Mr. Mayling's, for fear of being troublesome, persuaded her to go to her house at Rush Common, near Brixton Causeway, where she would endeavour to make her as happy as she could, hoping by this time all was over, as nothing had happened at that gentleman's house while she was there. This was about two o'clock in the afternoon.

" Mr. and Miss Gresham were at Mr. Pain's house, when Mrs. Pain, Mrs. Golding, and her maid went there. It being about dinner time, they all dined together; in the interim, Mrs. Golding's servant was sent to her house to see how things remained. When she returned, she told them nothing had happened since they left it. Some time after, Mr. Gresham and Miss went home, everything remaining

quiet at Mr. Pain's: but about eight o'clock
in the evening a fresh scene began ; the first
thing that happened, was a whole row of
pewter dishes, except one, fell from off a shelf
to the middle of the floor, rolled about a little
while, then settled; and, what is almost
beyond belief, as soon as they were quiet,
turned upside down ; they were then put on
the dresser, and went through the same a
second time ; next fell a whole row of pewter
plates from off the second shelf over the dresser
to the ground, and being taken up and put on
the dresser one in another, they were thrown
down again.

"The next thing was two eggs that were
upon one of the pewter shelves, one of them
flew off, crossed the kitchen, struck a cat on
the head, and then broke in pieces.

"Next, Mary Martin, Mrs. Pain's servant,
went to stir the kitchen fire, she got to the
right hand side of it, being a large chimney,
as is usual in farm houses, a pestle and mortar
that stood nearer the left hand end of the
chimney shelf, jumped about six feet on the
floor.   Then went candlesticks and other
brasses, scarce anything remaining in its place.
After this, the glasses and china were put down
on the floor for fear of undergoing the same

fate, they presently began to dance and tumble about, and then broke to pieces. A tea-pot that was among them, flew to Mrs. Golding's maid's foot, and struck it.

" A glass tumbler that was put on the floor, jumped about two feet and then broke. Another that stood by it jumped about at the same time, but did not break till some hours after, when it jumped again, and then broke. A china bowl that stood in the parlour jumped from the floor to behind a table that stood there. This was most astonishing, as the distance from where it stood was between seven and eight feet, but was not broke. It was put back by Richard Fowler to its place, where it remained some time, and then flew to pieces.

"The next thing that followed was a mustard pot, that jumped out of a closet and was broke. A single cup that stood upon the table (almost the only thing remaining) jumped up, flew across the kitchen, ringing like a bell, and then was dashed to pieces against the dresser. A candlestick that stood on the chimney-shelf flew across the kitchen to the parlour door, at about fifteen feet distance. A tea-kettle under the dresser, was thrown out about two feet; another kettle that stood at one end of the range, was thrown against the iron that is

fixed to prevent children falling into the fire.
A tumbler with rum and water in it, that stood
upon a waiter upon a table in the parlour,
jumped about ten feet, and was broke. The
table then fell down, and along with it a silver
tankard belonging to Mrs. Golding, the waiter
in which stood the tumbler and a candlestick.
A case bottle then flew to pieces.

"The next circumstance was a ham that
hung in one side of the kitchen chimney, it
raised itself from the hook and fell down to the
ground. Some time after, another ham that
hung on the other side of the chimney, likewise underwent the same fate. Then a flitch of
bacon which hung up in the same chimney fell
down.

"All the family were eye-witnesses to these
circumstances, as well as other persons, some of
whom were so alarmed and shocked, that they
could not bear to stay, and were happy in
getting away, though the unhappy family were
left in the midst of their distresses. Most of
the genteel families around were continually
sending to inquire after them, and whether
all was over or not. Is it not surprising that
some among them had not the inclination
and resolution to try to unravel this most
intricate affair, at a time when it would have

been in their power to have done so; there certainly was sufficient time for so doing, as the whole, from first to last, continued upwards of twenty hours.

" At all the times of action, Mrs. Golding's servant was walking backwards and forwards, either in the kitchen or parlour, or wherever some of the family happened to be. Nor could they get her to sit down five minutes together, except at one time for about half an hour towards the morning, when the family were at prayers in the parlour; then all was quiet: but in the midst of the greatest confusion, she was as much composed as at any other time, and with uncommon coolness of temper advised her mistress not to be alarmed or uneasy, as she said, these things could not be helped. Thus she argued, as if they were common occurrences which must happen in every family.

" This advice surprised and startled her mistress, almost as much as the circumstances that occasioned it. For how can we suppose that a girl of about twenty years old (an age when female timidity is too often assisted by superstition) could remain in the midst of such calamitous circumstances (except they proceed from causes best known to herself) and not be struck with the same terror as every other per-

son was who was present. These reflections led
Mr. Pain, and at the end of the transactions,
likewise Mrs. Golding, to think that she was
not altogether so unconcerned as she appeared
to be. But hitherto, the whole remains mysterious and unravelled.

"About ten o'clock at night, they sent over
the way to Richard Fowler, to desire he would
come and stay with them. He came and continued till one in the morning, and was so terrified that he could remain no longer.

"As Mrs. Golding could not be persuaded
to go to bed, Mrs. Pain at that time (one
o'clock) made an excuse to go up stairs to her
youngest child, under pretence of getting it to
sleep, but she really acknowledges it was
through fear, as she declares she could not sit
up to see such strange things going on, as
everything, one after another, was broke, till
there was not above two or three cups and
saucers remaining out of a considerable quantity of china, &c., which was destroyed to
the amount of some pounds.

"About five o'clock on Tuesday morning,
Mrs. Golding went up to her niece, and
desired her to get up, as the noises and
destruction were so great, she could continue
in the house no longer. At this time all the

tables, chairs, drawers, &c., were tumbling about. When Mrs. Pain came down, it was amazing beyond all description. Their only security then was to quit the house for fear of the same catastrophe, as had been expected the morning before, at Mrs. Golding's; in consequence of this resolution, Mrs. Golding and her maid went over the way to Richard Fowler's. When Mrs. Golding's maid had seen her safe to Richard Fowler's, she came back to Mrs. Pain, to help her to dress the children in the barn, where she had carried them for fear of the house falling. At this time all was quiet; they then went to Fowler's, and then began the same scene as had happened at the other places. It must be remarked, all was quiet here as well as elsewhere, till the maid returned.

"When they got to Mr. Fowler's, he began to light a fire in his back room. When done, he put the candle and candlestick upon a table in the fore room. This apartment Mrs. Golding and her maid had passed through. Another candlestick with a tin lamp in it, that stood by it, were both dashed together, and fell to the ground. A lanthorn with which Mrs. Golding was lighted with cross the road, sprung from a hook to the ground, and a

quantity of oil spilled on the floor.   The last
thing was the basket of coals tumbled over;
the coals rolling about the room; the maid
then desired Richard Fowler not to let her
mistress remain there, as she said, wherever
she was, the same things would follow.   In
consequence of this advice, and fearing greater
losses to himself, he desired she would quit
his house; but first begged her to consider
within herself, for her own and the public's
sake, whether or not she had not been guilty
of some atrocious crime, for which Providence
was determined to pursue her on this side of
the grave, for he could not help thinking, she
was the object that was to be made an example to posterity, by the All-seeing eye of
Providence, for crimes which but too often
none but that Providence can penetrate, and
by such means as these bring to light.

"Thus was the poor gentlewoman's measure of affliction complete, not only to have
undergone all which has been related, but to
have added to it the character of a bad and
wicked woman, when till this time, she was
esteemed as a most deserving person.   In
candour to Fowler, he could not be blamed;
what could he do? what would any man have
done that was so circumstanced?   Mrs·

Golding soon satisfied him ; she told him she would not stay in his house, or any other person's, as her conscience was quite clear, and she could as well wait the will of Providence in her own house as in any other place whatever; upon which she and her maid went home. Mr. Pain went with them. After they had got to Mrs. Golding's the last time, the same transactions once more began upon the remains that were left.

"A nine gallon cask of beer, that was in the cellar, the door being open, and no person near it, turned upside down. A pail of water that stood on the floor, boiled like a pot. A box of candles fell from a shelf in the kitchen to the floor; they rolled out, but none were broke: and a round mahogany table overset in the parlour.

"Mr. Pain then desired Mrs. Golding to send her maid for his wife to come to them; when she was gone, all was quiet; upon her return she was immediately discharged, and no disturbances have happened since; this was between six and seven o'clock on Tuesday morning.

"At Mrs. Golding's were broke the quantity of three pails-full of glass, china, &c. At Mrs. Pain's they filled two pails.

" Thus ends the narrative: a true, circumstantial, and faithful account of which we have
laid before the public; and have endeavoured
as much as possible, throughout the whole, to
state only facts, without presuming to obtrude
any opinion on them. If we have in part hinted
anything that may appear unfavourable to the
girl, it is not from a determination to charge
her with the cause, right or wrong, but only
from a strict adherence to truth, most sincerely
wishing this extraordinary affair may be unravelled.

"The above narrative is absolutely and
strictly true, in witness whereof we have set our
hands this eleventh day of January, 1772.

" MARY GOLDING

"JOHN PAIN

" MARY PAIN

" RICHARD FOWLER

" SARAH FOWLER

" MARY MARTIN.

" The original copy of this narrative, signed
as above, with the parties own hands, was put
into the hands of Mr. Marks, bookseller, in
St. Martin's Lane, to satisfy persons who
choose to inspect the same."

Such phenomena as this of the Stockwell
Ghost, are by no means uncommon, and I am

acquainted with many more instances than I can allude to here. One occurred very lately in the neighbourhood of London, as I learnt from the following newspaper paragraph. I subsequently heard that the little girl had been sent away, but whether the phenomena then ceased, or whether she carried the disturbance with her, I have not been able to ascertain, nor does it appear certain that she had anything to do with it :—

"A Mischievous and Mysterious Ghost. (From a Correspondent.)—The whole of the neighbourhood of Black Lion-lane, Bayswater, is ringing with the extraordinary occurrences that have recently happened in the house of a Mr. Williams, in the Moscow-road, and which bear a strong resemblance to the celebrated Stockwell ghost affair in 1772. The house is inhabited by Mr. and Mrs. Williams, a grown up son and daughter, and a little girl between ten and eleven years of age. On the first day, the family, who are remarkable for their piety, were startled all at once by a mysterious movement among the things in the sitting-rooms and kitchen, and other parts of the house. At one time, without any visible agency, one of the jugs came off the hook over the dresser, and was broken; then followed another, and

next day another. A china tea-pot, with the
tea just made in it, and placed on the mantel-
piece, whisked off on to the floor, and was
smashed. A pewter one, which had been sub-
stituted immediately after, did the same, and
when put on the table, was seen to hop about
as if bewitched, and was actually held down
while the tea was made for Mr. Williams's
breakfast, before leaving for his place of
business. When for a time all had been quiet,
off came from its place on the wall, a picture
in a heavy gilt frame, and fell to the floor
without being broken. All was now amazement and terror, for the old people are very
superstitious, and ascribing it to a supernatural
agency, the other pictures were removed, and
stowed away on the floor. But the spirit of
locomotion was not to be arrested. Jugs and
plates continued at intervals to quit their posts,
and skip off their hooks and shelves into the
middle of the room, as though they were
inspired by the magic flute, and at supper,
when the little girl's mug was filled with beer,
the mug slided off the table on to the
floor. Three times it was replenished and
replaced, and three times it moved off again.
It would be tedious to relate the fantastic tricks
which have been played by household articles

of every kind. An Egyptian vase jumped off the table suddenly when no soul was near, and was smashed to pieces. The tea-kettle popped off the fire into the grate as Mr. Williams had filled the teapot, which fell off the chimney-piece. Candlesticks, after a dance on the table, flew off, and ornaments from the shelves, and bonnets and cap-boxes flung about in the oddest manner. A looking-glass hopped off a dressing-table, followed by combs and brushes and several bottles, and a great pin-cushion has been remarkably conspicuous for its incessant jigs from one part to another. The little girl, who is a Spaniard, and under the care of Mr. and Mrs. Williams, is supposed by their friends to be the cause of it all, however extraordinary it may seem in one of her age, but up to the present time it continues a mystery, and the *modus operandi* is invisible."
—*Morning Post.*

To imagine that these extraordinary effects were produced by the voluntary agency of the child, furnishes one of those remarkable instances of the credulity of the sceptical, to which I have referred. But when we read a true statement of the effects involuntarily exhibited by Angelique Cottin, we begin to see that it is just possible the other strange phenomena may be provided by a similar agency.

The French Academy of Sciences had determined, as they had formerly done by Mesmerism, that the thing should not be true, and Monsieur Arago was nonsuited; but although it is extremely possible that either the phenomenon had run its course and arrived at a natural termination, or that the removal of the girl to Paris had extinguished it, there appears no doubt that it had previously existed.

Angelique Cottin was a native of La Perriere, aged fourteen, when on the 15th January, 1846, at eight o'clock in the evening, whilst weaving silk gloves at an oaken frame, in company with other girls, the frame began to jerk and they could not by any efforts keep it steady. It seemed as if it were alive, and becoming alarmed, they called in the neighbours, who would not believe them; but desired them to sit down and go on with their work. Being timid, they went one by one, and the frame remained still, till Angelique approached, when it recommenced its movements, whilst she was also attracted by the frame: thinking she was bewitched or pos-

sessed, her parents took• her to the Presbytery
that the spirit might be exorcised. The
curate however, being a sensible man, refused
to do it; but set himself, on the contrary, to
observe the phenomenon; and being perfectly
satisfied of the fact, he bade them take her
to a physician.

Meanwhile, the intensity of the influence,
whatever it was, augmented; not only articles
made of oak, but all sorts of things were acted
upon by it and reacted upon her, whilst.per-
sons who were near her, even without contact,
frequently felt electric shocks. The effects,
which were diminished when she was on a
carpet or even a waxed cloth, were most
remarkable when she was on the bare earth.
They sometimes entirely ceased for two or
three days, and then recommenced. Metals
were not affected. Anything touched by her
apron or dress would fly off, although a person
held it; and Monsieur Hebert whilst seated on
a heavy tub or trough, was raised up with it.
In short, the only place she could repose on,
was a stone covered with cork ; they also kept
her still by isolating her. When she was
fatigued the effects diminished. A needle
suspended horizontally, oscillated rapidly with
the motion of her arm without contact, or

remained fixed, whilst deviating from the magnetic direction. Great numbers of enlightened medical and scientific men witnessed these phenomena, and investigated them with every precaution to prevent imposition. She was often hurt by the violent involuntary movements he was thrown into, and was evidently afflicted by chorea.

Unfortunately, her parents poor and ignorant, insisted much against the advice of the doctors, on exhibiting her for money; and under these circumstances, she was brought to Paris; and nothing is more probable, than that after the phenomena had really ceased, the girl may have been induced to simulate, what had originally been genuine, the thing avowedly ceased altogether on the evening of the 10th April, and there has been no return of it.

In 1831, a young girl, also aged fourteen, who lived as under nursery-maid in a French family, exhibited the same phenomena; and when the case of Angelique Cottin was made public, her master published hers. He says that things of such an extraordinary nature occurred as he dare not repeat, since none but an eye-witness could believe them. The thing lasted for three years, and there was ample time for observation.

In the year 1686, a man at Brussels, called Breekmans was similarly effected. A commission was appointed by the magistrates to investigate his condition; and, being pronounced a sorcerer, he would have been burnt, had he not luckily made his escape.

Many somnambulic persons are capable of giving an electric shock; and I have met with one person, not somnambulic, who informs me that he has frequently been able to do it by an effort of the will.

Dr. Ennemoser relates the case of a Mademoiselle Emmerich, sister to the professor of theology at Strasburg, who also possessed this power. This young lady, who appears to have been a person of very rare merit and endowments, was afflicted with a long and singular malady, originating in a fright, in the course of which she exhibited many very curious phenomena, having fallen into a state of natural somnambulism, accompanied by a high degree of lucidity. Her body became so surcharged with electricity, that it was necessary to her relief to discharge it ; and she sometimes imparted a complete battery of shocks to her brother and her physician, or whoever was near, and that, frequently, when they did not touch her. Professor Emmerich mentions also, that she sent him a smart shock, one day, when he was several rooms off. He started up and rushed into her chamber, where she was in bed, and as soon as she saw him she said, laughing, "Ah, you felt it, did you?" Mademoiselle Emmerich's illness terminated in death.

Cotugno, a surgeon, relates that having touched with his scalpel, the intercostal nerve of a mouse that had bitten his leg, he received an electric shock; and where the torpedo abounds, the fishermen, in pouring water over the fish they have caught for the purpose of washing them, know if one is amongst them by the shock they sustain.

A very extraordinary circumstance, which we may possibly attribute to some such influence as the above occurred at Rambouillet, in November, 1846. The particulars are furnished by a gentleman residing on the spot at the time, and were published by the Baron Dupotel; who however attempts no explanation of the mystery.

One morning, some travelling merchants, or pedlars, came to the door of a farm house, belonging to a man named Bottel, and asked for some bread, which the maid servant gave them and they went away. Subsequently one of the party returned to ask for more and was refused. The man I believe expressed some resentment, and uttered vague threats, but she would not give him anything, and he departed. That night at supper the plates began to dance and to roll off the table, without any visible cause, and several other unaccountable phenomena occurred; and the girl going to the door and chancing to place herself just where the pedlar had stood, she was seized with convulsions and an extraordinary rotatory motion. The carter who was standing by, laughed at her, and out of bravado, placed himself on the same spot, when he felt almost suffocated, and was so unable to command his movements, that he was overturned into a large pool that was in front of the house.

Upon this, they rushed to the curé of the parish for assistance, but he had scarcely said a prayer or two, before he was attacked in the same manner, though in his own house, and his furniture beginning to oscillate and crack as if it were bewitched, the poor people were frightened out of their wits.

By and by the phenomena intermitted, and they hoped all was over; but presently it began again; and this occurred more than once before it subsided wholly.

VOL. II.   2 A

On the 8th December, 1836, at Stuttgard, Carl Fischer, a baker's hoy, aged seventeen, of steady habits and good character, was fixed with a basket on his shoulders in some unaccountable way in front of his master's house. He foresaw the thing was to happen when he went out very early, with his bread in the morning; earnestly wished that the day was over, and told his companion that if he could only cross the threshold, on his return, he should escape it. It was about six when he did return; and his master hearing a fearful noise, which he could not describe, " as if proceeding from a multitude of beings," looked out of the. window, where he saw Carl violently struggling and fighting with his apron, though his feet were immoveably fixed to one spot. A hissing sound proceeded from his mouth and nose, and a voice which was neither his nor that of any person present, was heard to cry, " Stand fast, Carl!" The master says, that he could not have believed such a thing ; and he was so alarmed that he did not venture into the street, where numerous persons were assembled. The boy said he must remain there till eleven o'clock: and the police kept guard over him till that time, as the physician said he must not be interfered with, and the people sought to push him from the spot.

When the time had expired, he was carried to the hospital, where he seemed exceedingly exhausted, and fell into a profound sleep.

I meet with numerous extraordinary records of a preternatural ringing of all the bells in a house; sometimes occurring periodically for a considerable time; and continuing after precautions have been taken which precluded the possibility of trick or deception, the wires being cut, and vigilant eyes watching them; and yet they rung on by day or night, just the same.

It is certainly very difficult to conceive, but at the same time it is not impossible, that such strange phenomena as that of the Stockwell Ghost and many similar ones, may be the manifestations of some extraordinary electrical influence; but there are other cases of poltergeist, which it is impossible to attribute to the same cause, since they are accompanied by evident manifestations of will and intelligence. Such was the instance related in Southey's life of Wesley, which occurred in the year 1716, beginning with a groaning, and subsequently proceeding to all manner of noises, lifting of latches, clattering of windows, knockings of a most mysterious kind, &c. &c. The family were not generally frightened, but the young children, when asleep, showed symptoms of great terror. This annoyance lasted, I think, two or three months, and then ceased. As in most of these cases, the dog was extremely frightened, and hid himself when the visitations commenced.

In the year 1838, a circumstance of the same kind occurred in Paris, in the Rue St. Honoré, and not very along ago, there was one in Caithness, in which most unaccountable circumstances transpired. Amongst the rest, stones were flung, which never hit people, but fell at their feet, in rooms perfectly closed on all sides. A gentleman who witnessed these extraordinary phenomena, related the whole story to an advocate of my acquaintance ; who assured me, that however impossible he found it to credit such things, he should certainly place entire reliance on that gentleman's word in any other case.

Then there is the famous story of the Drummer of Tedworth;* and the persecution of Professor Schuppart, at Giessen, in Upper Hesse, which continued with occasional intermission for six years. This affair began with

* There was also a remarkable case of this sort at Mr. Chaves, in Devonshire, in the year 1810, where affidavits were made before the magistrates attesting the facts, and large rewards offered for discovery; but in vain. The phenomena continued several months, and the spiritual agent was freqnently seen in the form of some strange animal.

a violent knocking at the door one night; next
day stones were sent whizzing through closed
rooms in all directions; so that although no
one was struck, the windows were all broken;
and no sooner were new panes put in, than
they were broken again. He was persecuted
with slaps on the face by day and by night, so
that he could get no rest; and when two persons were appointed by the authorities to sit
by his bed to watch him, they got the slaps
also. When he was reading at his desk, his
lamp would suddenly rise up and remove to
the other end of the room—not as if thrown,
but evidently carried: his books were torn to
pieces and flung at his feet, and when he was
lecturing, this mischievous sprite would tear
out the leaf he was reading; and it is very remarkable that the only thing that seemed
available, as a protection, was a drawn sword
brandished over his head by himself, or others,
which was one of the singularities attending
the case of the Drummer of Tedworth. Schuppart narrated all these circumstances in his public lectures, and nobody ever disputed the facts.

A remarkable case of this sort occurred in
the year 1670, at Keppock, near Glasgow;
there also stones were thrown which hit nobody; but the annoyance only continued eight

days; and there are several more to be found
recorded in works of that period. The disturbance that happened in the house of Gilbert
Cambell, at Glenluce, excited considerable
notice. Here, as elsewhere, stones were thrown;
but as in most similar instances I meet with,
no human being was damaged; the licence
of these spirits or goblins, or whatever they
be, seeming to extend no further than worry-
ing and tormenting their victims. In this
case, however, the spirit spoke to them, though
he was never seen. The annoyance commenced in November, of the year 1654, I think,
and continued till April, when there was some
intermission till July, when it recommenced.
The loss of the family from the things destroyed was ruining; for their household goods
and chattels were rendered useless, their food
was polluted and spoiled, and their very clothes
cut to pieces whilst on their backs by invisible
hands; and it was in vain that all the ministers
about the country assembled to exorcise this
troublsesome spirit, for whoever was there the
thing continued exactly the same.

At length, poor Cambell applied to the
Synod of Presbyters for advice, and a meeting
was convened in October, 1655, and a solemn
day of humiliation was imposed through the
whole bounds of the Presbytery, for the sake

of the afflicted family. Whether it was owing
to this or not, there ensued an alleviation from
that time till April; and from April till August
they were entirely free, and hoped all was over;
but then it began again worse than ever, and
they were dreadfully tormented through the
autumn; after which the disturbance ceased,
and although the family lived in the house
many years afterwards, nothing of the sort
ever happened again.

There was another famous case, which
occurred at a place called Ring-Croft, in Kirk-
cudbright, in the year 1695. The afflicted
family bore the name of Mackie. In this
instance, the stones did sometimes hit them,
and they were beaten as if by staves; they, as
well as strangers who came to the house,
were lifted off the ground by their clothes,
their bed coverings were taken off their beds;
things were visibly carried about the house
by *in*visible hands; several people were hurt,
even to the effusion of blood, by stones and
blows; there were fire balls seen about the
house, which was several times actually
ignited; people, both of the family and others,
felt themselves grasped as if by a hand; then
there was groaning, crying, whistling, and a
voice that frequently spoke to them; crowds
of people went to the house, but the thing

continued just the same whether there were many or few, and sometimes the whole building shook as if it were coming down.

A day of humiliation was appointed in this case also, but without the least effect. The disturbance commenced in February, and ended on the 1st of May. Numberless people witnessed the phenomena, and the account of it is attested by fourteen ministers and gentlemen.

The same sort of thing occurred in the year 1659, in a place inhabited by an Evangelical bishop, called Schlotterbeck. It began in the same manner by throwing of stones and other things, many of which came through the roof; insomuch that they believed at first that some animal was concealed there. However, nothing could be found, and the invisible guest soon proceeded to other annoyances similar to those above-mentioned; and though they could not see him, his footsteps were for ever heard about the house. At length, wearied out, the bishop applied to the Government for aid, and they sent him a company of soldiers to guard the house by day and night, out of which he and his family retired. But the goblin cared no more for the soldiers than it had done for the city watch; the thing continued without intermission, whoever was there, till it ceased of its own accord. There was a house at Aix la Chapelle, that was for several years quite uninhabitable from a similar cause.

I could mention many other cases, and, as I have said before, they occur in all countries, but these will suffice as specimens of the class. It is in vain for people who were not on the spot to laugh, and assert that these were the mischievous tricks of servants, or others, when those who were there, and who had such a deep interest in unravelling the mystery, and such abundance of time and opportunity for doing it, could find no solution whatever. In many of the above cases, the cattle were unloosed, the horses were turned out of their stables, and uniformly all the animals, in the way, exhibited great terror, sweating and trembling whilst the visitation continued.

Since we cannot but believe that man forms but one class in an immense range of existences, do not these strange occurrences suggest the idea, that occasionally some individual out of this gamut of beings comes into rapport with us, or crosses our path like a comet, and that, whilst certain conditions last, it can hover about us, and play these *puckish*, mischievous tricks, till the charm is broken, and then it re-enters its own sphere, and we are cognizant of it no more !

But one of the most extraordinary examples of this kind of annoyance, is that which occurred in the year 1806, in the castle of Prince Hohenlohe, in Silesia. The account is given at length by Councillor Hahn, of Ingelfingen, who witnessed the circumstances ; and, in consequence of the various remarks that have been since made on the subject, in different publications, he has repeatedly re-asserted the facts in letters which have been printed and laid before the public. I cannot, therefore, see what right we have to disbelieve a man of honour and character, as he is said to be, merely because the circumstances he narrates are unaccountable, more especially as the story, strange as it is, by no means stands alone in the annals of demonology. The following details were written down at the time the events occurred, and they were communicated by Councillor Hahn to Dr. Kerner in the year 1828.

" After the campaign of the Prussians against the French in the year 1806, the reigning Prince of Hohenlohe gave orders to Councillor Hahn, who was in his service, to proceed to Slawensick, and there to wait his return.

His Serene Highness advanced from Leignitz
towards his principality, and Hahn also commenced his journey towards Upper Silesia on
the 19th November. At the same period,
Charles Kern, of Kuntzlau, who had fallen
into the hands of the French, being released on
parole, and arriving at Leignitz, in an infirm
condition, he was allowed to spend some time
with Hahn, whilst awaiting his exchange.

" Hahn and Kern had been friends in their
youth, and their destinies having brought them
both at this time into the Prussian States, they
were lodged together in the same apartment of
the castle, which was one on the first floor, forming an angle at the back of the building, one
side looking towards the north, and the other to
the east. On the right of the door of this room
was a glass door, which led into a chamber
divided from those which followed by a wainscoat partition. The door in this wainscoat,
which communicated to those adjoining rooms,
was entirely closed up, because in them all sorts
of household utensils were kept. Neither in
this chamber, nor in the sitting-room which
preceded it, was there any opening whatever
which could furnish the means of communication from without; nor was there any body
in the castle besides the two friends, except

the Prince's two coachmen and Hahn's servant.
The whole party were fearless people; and as
for Hahn and Kern, they believed in nothing
less than ghosts or witches, nor had any previous experience induced them to turn their
thoughts in that direction. Hahn, during his
collegiate life, had been much given to philosophy—had listened to Fichte, and earnestly
studied the writings of Kant. The result of
his reflections was a pure materialism; and he
looked upon created man, not as an aim, but
merely as a means to a yet undeveloped end.
These opinions he has since changed, like
many others who think very differently in
their fortieth year to what they did in their
twentieth. The particulars here given are
necessary in order to obtain credence for the
following extraordinary narrative; and to
establish the fact that the phenomena were
not merely accepted by ignorant superstition,
but coolly and courageously investigated by
enlightened minds. During the first days of
their residence in the castle, the two friends,
living together in solitude, amused their long
evenings with the works of Schiller, of whom
they were both great admirers; and Hahn
usually read aloud. Three days had thus
passed quietly away, when, as they were sit-

ting at the table, which stood in the middle of
the room, about nine o'clock in the evening,
their reading was interrupted by a small
shower of lime, which fell around them. They
looked at the ceiling, concluding it must have
come thence, but could perceive no abraded
parts; and whilst they were yet seeking to
ascertain whence the lime had proceeded,
there suddenly fell several larger pieces, which
were quite cold, and appeared as if they had
belonged to the external wall. At length
concluding the lime must have fallen from
some part of the wall, and giving up further
enquiry, they went to bed, and slept quietly
till morning, when, on awaking, they were
somewhat surprised at the quantity which
strewed the floor, more especially as they
could still discover no part of the walls or
ceiling from which it could have fallen. But
they thought no more of the matter till evening, when, instead of the lime falling as before,
it was thrown, and several pieces struck Hahn;
At the same time they heard heavy blows,
sometimes below, and sometimes over their
heads, like the sound of distant guns; still
attributing these sounds to natural causes,
they went to bed as usual, but the uproar
prevented their sleeping, and each accused the

other of occasioning it by kicking with his feet
against the foot-board of his bed, till, finding
that the noise continued when they both got
out and stood together in the middle of the
room, they were satisfied that this was not the
case. On the following evening, a third noise
was added, which resembled the faint and distant beating of a drum. Upon this, they
requested the governess of the castle to send
them the key of the apartments above and
below, which was brought them by her son;
and, whilst he and Kern went to make their
investigations, Hahn remained in their own
room. Above, they found an empty] room;
below, a kitchen. They knocked, but the
noise they made was very different to that
which Hahn continued all the while to hear
around him. When they returned, Hahn said
jestingly, 'The place is haunted!' On this
night, when they went to bed with a light
burning, they heard what seemed like a person walking about the room with slippers on,
and a stick, with which he struck the floor as
he moved step by step. Hahn continued to
jest, and Kern to laugh, at the oddness of these
circumstances for some time, when they both,
as usual, fell asleep, neither in the slightest
degree disturbed by these events, nor inclined

to attribute them to any supernatural cause. But on the following evening the affair became more inexplicable; various articles in the room were thrown about; knives, forks, brushes, caps, slippers, padlocks, funnel, snuffers, soap —everything, in short, that was moveable; whilst lights darted from corner, and everything was in confusion; at the same time the lime fell, and the blows continued. Upon this, the two friends called up the servants, Knittel, the castle watch, and whoever else was at hand, to be witnesses of these mysterious operations. In the morning all was quiet, and generally continued so till about an hour after midnight. One evening, Kern going into the above-mentioned chamber to fetch something, and hearing such an uproar that it almost drove him backwards to the door, Hahn caught up the light, and both rushed into the room, where they found a large piece of wood, lying close to the wainscoat. But supposing this to be the cause of the noise, who had set it in motion? For Kern was sure the door was shut, even whilst the noise was making; neither had there been any wood in the room. Frequently, before their eyes, the knives and snuffers rose from the table, and fell, after some minutes, to the ground; and Hahn's large shears were once

lifted in this manner between him and one of
the Prince's cooks, and, falling to the ground,
stuck into the floor.   As some nights, however,
passed quite quietly, Hahn was determined not
to leave the rooms; but when, for three weeks,
the disturbance was so constant that they
could get no rest, they resolved on removing
their beds into the large room above, in
hopes of once more enjoying a little quiet
sleep.   Their hopes were vain—the thumping continued as before; and not only so,
but articles flew about the room, which they
were quite sure they had left below.   'They
may fling as they will,' cried Hahn, 'sleep I
must;' whilst Kern began to undress, pon-
dering on these matters as he walked up and
down the room.   Suddenly Hahn saw him
stand, as if transfixed, before the looking glass,
on which he had accidentally cast his eyes.   He
had so stood for some minutes, when he was
seized with a violent trembling, and turned
from the mirror with his face as white as
death.   Hahn, fancying the cold of the uninhabited room had seized him, hastened to
throw a cloak over him; when Kern, who was
naturally very courageous, recovered himself,
and related, though with trembling lips, that,
as he had accidentally looked in the glass, he

had seen a white female figure looking out of
it; she was in front of his own image, which
he distinctly saw behind her. At first he
could not believe his eyes; he thought it must
be fancy, and for that reason he had stood so
long; but when he saw that the eyes of the
figure moved, and looked into his, a shudder had seized him, and he had turned away.
Hahn upon this advanced with firm steps to
the front of the mirror, and called upon the
apparition to show itself to him; but he saw
nothing, although he remained a quarter of an
hour before the glass, and frequently repeated
his exhortation. Kern then further related
that the features of the apparition were very
old, but not gloomy or morose; the expression
indeed was rather that of indifference; but the
face was very pale, and the head was wrapped
in a cloth which left only the features visible.

"By this time it was four o'clock in the morning—sleep was banished from their eyes,—and
they resolved to return to the lower room, and
have their beds brought back again; but the
people who were sent to fetch them returned,
declaring they could not open the door, although it did not appear to be fastened. They
were sent back again; but a second and a
third time they returned with the same answer

Then Hahn went himself, and opened it with
the greatest ease. The four servants, however, solemnly declared, that all their united
strengths could make no impression on it.

"In this way a month had elapsed: the strange
events at the castle had got spread abroad ;
and amongst others who desired to convince
themselves of the facts, were two Bavarian
officers of dragoons, namely, Captain Cornet
and Lieutenant Magerle, of the regiment of
Minuci. Magerle offering to remain in the
room alone, the others left him, but scarcely
had they passed into the next apartment, when
they heard Magerle storming like a man in a
passion, and cutting away at the tables and
chairs with his sabre, whereupon the Captain
thought it advisable to return, in order to
rescue the furniture from his rage. They
found the door shut, but he opened it on their
summons, and related, in great excitement,
that as soon as they had quitted the room,
some cursed thing had begun to fling lime, and
other matters, at him ; and, having examined
every part of the room without being able to
discover the agent of the mischief, he had fallen
into a rage and cut madly about him.

"The party now passed the rest of the
evening together in the room, and the two

Bavarians closely watched Hahn and Kern, in order to satisfy themselves that the mystery was no trick of theirs. All at once, as they were quietly sitting at the table, the snuffers rose into the air, and fell again to the ground, behind Magerle; and a leaden ball flew at Hahn, and hit him upon the breast, and presently afterwards they heard a noise. at the glass door, as if somebody had struck his fist through it, together with a sound of falling glass. On investigation, they found the door entire, but a broken drinking-glass on the floor. By this time the Bavarians were convinced, and they retired from the room to seek repose in one more peaceful.

" Amongst other strange circumstances, the following, which occurred to Hahn is remarkable. One evening, about eight o'clock, being about to shave himself, the implements for the purpose, which were lying on a pyramidal stand in a corner of the room, flew at him, one after the other—the soap-box, the razor, the brush and the soap—and fell at his feet, although he was standing several paces from the pyramid. He and Kern, who was sitting at the table, laughed, for they were now so accustomed to these events that they only made them subjects of diversion. In the mean

time, Hahn poured some water, which had
been standing on the stove, in a basin, observing as he dipped his finger into it, that it was
of a nice heat for shaving. He seated himself
before the table, and strapped his razor; but
when he attempted to prepare the lather, the
water had clean vanished out of the basin.
Another time, Hahn was awakened by goblins
throwing at him a squeezed-up piece of sheet-
lead, in which tobacco had been wrapped; and
when he stooped to pick it up, the self-same
piece was flung at him again. When this was
repeated a third time, Hahn flung a heavy
stick at his invisible assailant.

"Dorfel, the book-keeper, was frequently a
witness to these strange events. He once laid
his cap on the table by the stove; when, being
about to depart, he sought for it, it had vanished. Four or five times he examined the
table in vain; presently afterwards he saw it
lying exactly where he had placed it when he
came in. On the same table, Knittel having
once placed his cap, and drawn himself a seat,
suddenly—although there was nobody near the
table—he saw it flying through the room to
his feet, where it fell.

"Hahn now determined to find out the
secret himself; and for this purpose seated him-

self, with two lights before him, in a position where he could see the whole of the room, and all the windows and doors it contained; but the same things occurred even when Kern was out, the servants in the stables, and nobody in the castle but himself; and the snuffers were as usual flung about, although the closest observation could not detect by whom.

"The forest-master, Radzensky, spent a night in the room; but although the two friends slept, he could get no rest. He was bombarded without intermission; and in the morning, his bed was found full of all manner of household articles.

"One evening, in spite of all the drumming and flinging, Hahn was determined to sleep; but a heavy blow on the wall, close to his bed, soon waked him from his slumbers. A second time he went to sleep, and was awaked by a sensation, as if some person had dipped his finger in water, and was sprinkling his face with it. He pretended to sleep again, whilst he watched Kern and Knittel, who were sitting at the table, the sensation of sprinkling returned; but he could find no water on his face.

"About this time, Hahn had occasion to make a journey as far as Breslau; and when he returned he heard the strangest story of all-

In order not to be alone in this mysterious
chamber, Kern had engaged Hahn's servant, a
man of about forty years of age, and of entire
singleness of character, to stay with him. One
night as Kern lay in his bed, and this man was
standing near the glass door in conversation
with him, to his utter amazement he beheld a
jug of beer, which stood on a table, in the
room, at some distance from him, slowly lifted
to a height of about three feet, and the contents poured into a glass, that was standing
there also, until the latter was half full. The
jug was then gently replaced, and the glass
lifted and emptied, as by some one drinking;
whilst John, the servant, exclaimed, in terrified
surprise, ' Lord Jesus! it swallows !' The
glass was quietly replaced, and not a drop of
beer was to be found on the floor. Hahn was
about to require an oath of John, in confirmation of this fact; but forbore, seeing how
ready the man was to take one, and satisfied of
the truth of the relation.

" One night Knetsch, an inspector of the
works, passed the night with the two friends,
and, in spite of the unintermitting flinging, they
all three went to bed. There were lights in
the room, and presently all three saw two
napkins, in the middle of the room, rise

slowly up to the ceiling, and, having there spread themselves out, flutter down again. The China bowl of a pipe, belonging to Kern, flew about and was broken. Knives and forks were flung; and at last one of the latter fell on Hahn's head, though, fortunately, with the handle downwards; and, having now endured this annoyance for two months, it was unanimously resolved to abandon this mysterious chamber, for this night at all events. John and Kern took up one of the beds, and carried it into the opposite room, but they were no sooner gone than a pitcher for holding chalybeate-water flew to the feet of the two who remained behind, although no door was open, and a brass candlestick was flung to the ground. In the opposite room the night passed quietly, although some sounds still issued from the forsaken chamber. After this, there was a cessation to these strange proceedings, and nothing more remarkable occurred, with the exception of the following circumstance. Some weeks after the above-mentioned removal, as Hahn was returning home, and crossing the bridge that leads to the castle gate, he heard the foot of a dog behind him. He looked round, and called repeatedly on the name of a greyhound that was much attached to him, thinking it might be she, but, although he still heard the foot, even when he ascended the stairs, as he could see nothing, he concluded it was an illusion. Scarcely, however, had he set his foot within the room, than Kern advanced and took the door out of his hand, at the same time calling the dog by name; adding, however, immediately that he thought he had seen the dog, but that he had no sooner called her than she disappeared. Hahn then inquired, if he had really seen the dog. 'Certainly I did,' replied Kern; "she was close behind you—half within the door—and that was the reason I took it out of your hand, lest, not observing her you should have shut it suddenly, and crushed her. It was a white dog, and I took it for Flora.' Search was immediately made for the dog, but she was found locked up in the stable, and had not been out of it the whole day. It is certainly remarkable—even supposing Hahn to have been deceived with respect to the footsteps—that Kern should have seen a white dog behind him, before he had heard a word on the subject from his friend, especially as there was no such animal in the neighbourhood; besides, it was not yet dark, and Kern was very sharp-sighted.

" Hahn remained in the castle for half-a-year after this, without experiencing anything extraordinary; and even persons who had possession of the mysterious chambers, were not subjected to any annoyance.

" The riddle, however, in spite of all the perquisitions and investigations that were set on foot remained unsolved—no explanation of these strange events could be found; and even supposing any motive could exist, there was nobody in the neighbourhood clever enough to have carried on such a system of persecution, which lasted so long, that the inhabitants of the chamber became almost indifferent to it.

" In conclusion, it is only necessary to add, that Councillor Hahn wrote down this account for his own satisfaction, with the strictest regard to truth. His words are :—

" ' I have described these events exactly as I heard and saw them; from beginning to end I observed them with the most entire self-possession. I had no fear, nor the slightest tendency to it; yet the whole thing remains to me perfectly inexplicable. Written the 19th November, 1808.

" ' Augustus Hahn, Councillor.'
" Doubtless many natural explanations of

these phenomena will be suggested, by those who consider themselves above the weakness of crediting stories of this description. Some say that Kern was a dextrous juggler, who contrived to throw dust in the eyes of his friend Hahn; whilst others affirm that both Hahn and Kern were intoxicated every evening. I did not fail to communicate these objections to Hahn, and here insert his answer.

"'After the events alluded to, I resided with Kern for a quarter of a year in another part of the Castle of Slawensick (which has been since struck by lightning, and burnt), without finding a solution of the mystery, or experiencing a repetition of the annoyance, which discontinued from the moment we quitted those particular apartments. Those persons must suppose me very weak, who can imagine it possible, that with only one companion, I could have been the subject of his sport for two months without detecting him. As for Kern himself, he was, from the first, very anxious to leave the rooms; but as I was unwilling to resign the hope of discovering some natural cause for these phenomena, I persisted in remaining; and the thing that at last induced me to yield to his wishes was his vexation at the loss of his China pipe, which had been flung against the wall and broken. Besides, jugglery requires a juggler, and I was frequently quite alone when these events occurred. It is equally absurd to accuse us of intoxication. The wine there was too dear for us to drink at all; and we confined ourselves wholly to weak beer. All the circumstances that happened are not set down in the narration; but my recollection of the whole is as vivid as if it had occurred yesterday. We had also many witnesses, some of whom have been mentioned. Councillor Klenk also visited me at a later period, with every desire to investigate the mystery; and when, one morning, he had mounted on a table, for the purpose of doing so, and was knocking at the ceiling with a stick, a powder horn fell upon him, which he had just before left on the table in another room. At that time Kern had been for some time absent. I neglected no possible means that could have led to a discovery of the secret; and at least as many people have blamed me, for my unwillingness to believe in a supernatural cause as the reverse. Fear is not my failing, as all who are acquainted with me know; and to avoid the possibility of error, I frequently asked others what they saw when

I was myself present; and their answers
always coincided with what I saw myself.
From 1809 to 1811 I lived in Jacobswald,
very near the castle where the Prince himself
was residing. I am aware that some singular
circumstances occurred whilst he was there;
but as I did not witness them myself, I cannot
speak of them more particularly.

" ' I am still as unable as ever to account
for those events, and I am content to submit
to the hasty remarks of the world, knowing
that I have only related the truth, and what
many persons now alive witnessed, as well as
myself.   " 'Councillor Hahn.

" ' Ingelfinger, 24th August, 1828.*' "

The only key to this mystery ever discovered
was, that after the destruction of the castle
by lightning, when the ruins were removed,
there was found the skeleton of a man without
a coffin. His skull had been split, and a
sword lay by his side.

Now, I am very well aware how absurd and
impossible these events will appear to many
people, and that they will have recourse to any
explanation rather than admit them for facts.
Yet, so late as the year 1835, a suit was brought
before the Sheriff of Edinburgh, in which Cap-

* Translated from the original German.—C. C.

tain Molesworth was defendant, and the landlord of the house he inhabited (which was at
Trinity, about a couple of miles from Edinburgh)
was plaintiff, founded upon circumstances not
so varied, certainly, but quite as inexplicable.
The suit lasted two years, and I have been
favoured with the particulars of the case by
Mr. M. L., the advocate employed by the
plaintiff, who spent many hours in examining
the numerous witnesses, several of whom
were officers of the army, and gentlemen of
undoubted honour and capacity for observation.

Captain Molesworth took the house of a
Mr. Webster, who resided in the adjoining one,
in May or June, 1835; and when he had been
in it about two months, he began to complain
of sundry extraordinary noises, which, finding
it impossible to account for, he took it into his
head, strangely enough, were made by Mr.
Webster. The latter naturally represented
that it was not probable he should desire to
damage the reputation of his own house, or
drive his tenant out of it, and retorted the
accusation. Still, as these noises and knockings continued, Captain M. not only lifted the
boards in the room most infected, but actually
made holes in the wall which divided his resi-

2 c 5

dence from Mr. W.'s, for the purpose of detecting the delinquent—of course without success. Do what they would, the thing went on just the same; footsteps of invisible feet, knockings, and scratchings, and rustlings, first on one side, and then on the other, were heard daily and nightly. Sometimes this unseen agent seemed to be knocking to a certain tune, and if a question were addressed to it which could be answered numerically, as, "How many people are there in this room?" for example, it would answer by so many knocks. The beds, too, were occasionally heaved up, as if somebody were underneath, and where the knockings were, the wall trembled visibly, but, search as they would, no one could be found. Captain Molesworth had had two daughters, one of whom, named Matilda, had lately died; the other, a girl between twelve and thirteen, called Jane, was sickly, and generally kept her bed; and, as it was observed that, wherever she was, these noises most frequently prevailed, Mr. Webster, who did not like the *mala fama* that was attaching itself to his house, declared that she made them, whilst the people in the neighbourhood believed that it was the ghost of Matilda, warning her sister that she was soon to follow.

Sheriff's officers, masons, justices of peace, and the officers of the regiment quartered at Leith, who were friends of Captain M., all came to his aid, in hopes of detecting or frightening away his tormenter, but in vain. Sometimes it was said to be a trick of somebody outside the house, and then they formed a cordon round it; and next, as the poor sick girl was suspected, they tied her up in a bag, but it was all to no purpose.

At length, ill and wearied out by the annoyances and the anxieties attending the affair, Captain M. quitted the house, and Mr. W brought an action against him for the damages committed, by lifting the boards, breaking the walls, and firing at the wainscoat, as well as for the injury done to his house by saying it was haunted, which prevented other tenants taking it.

The poor young lady died, hastened out of the world, it is said, by the severe measures used whilst she was under suspicion; and the persons that have since inhabited the house have experienced no repetition of the annoyance.

The manner in which these strange persecutions attach themselves to certain persons and places, seems somewhat analogous to

another class of cases, which bear a great
similarity to what was formerly called *possession;* and I must here observe, that many
German physicians maintain, that to this day
instances of genuine possession occur, and there
are several works published in their language
on the subject; and for this malady they consider magnetism the only remedy, all others
being worse than useless. Indeed, they look
upon *possession* itself as a demono-magnetic
state, in which the patient is in rapport with
mischievous or evil spirits; as in the Agatho
(or good) magnetic state, which is the opposite pole, he is in rapport with good ones; and
they particularly warn their readers against
confounding this infliction with cases of epi-
lepsy or mania. They assert that although instances are comparatively rare, both sexes and
all ages are equally subject to this misfortune;
and that it is quite an error to suppose, either,
that it has ceased since the Resurrection of
Christ, or that the expression used in the
Scriptures, "possessed by a devil," meant
merely insanity or convulsions. This disease,
which is not contagious, was well known to
the Greeks; and in later times Hofman has
recorded several well established instances.
Amongst the distinguishing symptoms, they

reckon the patient's speaking in a voice that is not his own, frightful convulsions and motions of the body, which arise suddenly, without any previous indisposition—blasphemous and obscene talk, a knowledge of what is secret, and of the future—a vomiting of extraordinary things, such as hair, stones, pins, needles, &c. &c. I need scarcely observe that this opinion is not universal in Germany; still, it obtains amongst many who have had considerable opportunities for observation.

Dr. Bardili had a case in the year 1830, which he considered decidedly to be one of possession. The patient was a peasant woman, aged thirty-four, who never had any sickness whatever; and the whole of whose bodily functions continued perfectly regular whilst she exhibited the following strange phenomena. I must observe that she was happily married, had three children ; was not a fanatic, and bore an excellent character for regularity and industry, when, without any warning or perceptible cause, she was seized with the most extraordinary convulsions, whilst a strange voice proceeded from her, which assumed to be that of an unblessed spirit, who had formerly inhabited a human form. Whilst these fits were on her, she entirely lost her own

individuality, and became this person; on returning to herself, her understanding and
character were as entire as before. The blas-
phemy and cursing, and barking and screech-
ing, were dreadful. She was wounded and injured severely by the violent falls and blows
she gave herself; and when she had an intermission, she could do nothing but weep over
what they told her had passed, and the state in
which she saw herself. She was moreover
reduced to a skeleton; for when she wanted
to eat, the spoon was turned round in her
hand, and she often fasted for days together.
This affliction lasted for three years; all reme-
dies failed, and the only alleviation she obtained
was by the continued and earnest prayers of
those about her and her own; for although
this demon did not like prayers, and violently
opposed her kneeling down, even forcing her
to outrageous fits of laughter, still they had a
power over him. It is remarkable that preg-
nancy, confinement, and the nursing her child,
made not the least difference in this woman's
condition. All went on regularly, but the
demon kept his post. At length, being mag-
netised, the patient fell into a partially somnambulic state, in which another voice was
heard to proceed from her, being that of her

protecting spirit, which encouraged her to patience and hope, and promised that the evil guest would be obliged to vacate his quarters. She often now fell into a magnetic state without the aid of a magnetiser. At the end of three years she was entirely relieved, and as well as ever.

In the case of Rosina Wildin, aged ten years, which occurred at Pleidelsheim, in 1834, the demon used to announce himself by crying out, " Here I am again!" Whereupon the weak, exhausted child, who had been lying like one dead, would rage and storm in a voice like a man's, perform the most extraordinary move-ments and feats of violence and strength, till he would cry out " Now I must be off again!" This spirit spoke generally in the plural number, for he said, she had another besides himself, a dumb devil, who plagued her most. " He it is that twirls her round and round, dis-torts her features, turns her eyes, locks her teeth, &c. What he bids me, I must do !" This child was at length cured by magnetism.

Barbara Rieger, of Steinbach, aged ten, in 1834, was possessed by two spirits, who spoke in two distinctly different male voices and dialects; one said he had formerly been a mason, the other gave himself out for a de-

ceased provisor; the latter of whom was much
the worst of the two. When they spoke, the
child closed her eyes, and when she opened
them again, she knew nothing of what they
had said. The mason confessed to have been
a great sinner, but the provisor was proud and
hardened, and would confess nothing. They
often commanded food, and made her eat it,
which, when she recovered her individuality,
she felt nothing of, but was very hungry. The
mason was very fond of brandy, and drank a
great deal; and if not brought when he
ordered it, his raging and storming was
dreadful. In her own individuality, the child
had the greatest aversion to this liquor. They
treated her for worms, and other disorders,
without the least effect; till at length, by
magnetism, the mason was cast out. The provisor was more tenacious, but, finally, they got
rid of him, too, and the girl remained quite
well.

In 1835, a respectable citizen, whose full
name in not given, was brought to Dr. Kerner.
He was aged thirty-seven, and till the last
seven years had been unexceptionable in conduct and character. An unaccountable change
had, however, come over him in his thirtieth
year, which made his family very unhappy;

and at length, one day, a strange voice suddenly spoke out of him, saying that he was the late magistrate, S., and that he had been in him for six years. When this spirit was driven out, by magnetism, the man fell to the earth, and was almost torn to pieces by the violence of the struggle; he then lay for a space as if dead, and arose quite well and free.

In another case, a young woman at Gruppenbach, was quite in her senses and heard the voice of her demon (who was also a deceased person), speak out of her, without having any power to suppress it.

In short, instances of this description seem by no means rare; and if such a phenomenon as possession ever did exist, I do not see what right we have to assert that it exists no longer, since, in fact, we know nothing about it; only, that being determined to admit nothing so contrary to the ideas of the present day, we set out by deciding that the thing is impossible.

Since these cases occur in other countries, no doubt they must do so in this; and, indeed, I have met with one instance much more remarkable in its details than any of those abovementioned, which occurred at Bishopwearmouth, near Sunderland, in the year 1840; and as the particulars of this case have been published and attested by two physicians and two surgeons, not to mention the evidence of numerous other persons, I think we are bound to accept the facts, whatever interpretation we may choose to put upon them.

The patient, named Mary Jobson, was between twelve and thirteen years of age; her parents, respectable people in humble life, and herself an attendant on a Sunday school. She became ill in November, 1839, and was soon afterwards seized with terrific fits, which continued, at intervals, for eleven weeks. It was during this period that the family first observed a strange knocking, which they could not account for. It was sometimes in one place, and sometimes in another; and even about the bed, when the girl lay in a quiet sleep, with her hands folded outside the clothes. They next heard a strange voice, which told them circumstances they did not know, but which they afterwards found to be correct. Then there was a noise like the clashing of arms, and such a rumbling that the tenant below thought the house was coming down; footsteps where nobody was to be seen, water falling on the floor, no one knew whence, locked doors opened, and above all, sounds of ineffably sweet music. The doctors and

the father were suspicious, and every precaution was taken, but no solution of the
mystery could be found. This spirit however
was a good one, and it preached to them, and
gave them a great deal of good advice. Many
persons went to witness this strange phenomenon, and some were desired to go by the
voice, when in their own homes. Thus
Elizabeth Gauntlett, whilst attending to some
domestic affairs at home, was startled by hearing a voice say, " Be thou faithful, and thou
shalt see the works of thy God, and shalt hear
with thine ears !" She cried out, " My God !
what can this be !" and presently she saw a
large white cloud near her. On the same
evening, the voice said to her, " Mary Jobson,
one of your scholars is sick; go and see her;
and it will be good for you." This person did
not know where the child lived ; but having
enquired the address, she went : and at the
door she heard the same voice bid her go up.
On entering the room, she heard another voice,
soft and beautiful, which bade her be faithful,
and said, " I am the Virgin Mary." This voice
promised her a sign at home ; and accordingly
that night, whilst reading the Bible, she heard
it say, " Jemima, be not afraid; it is I: if you
keep my commandments, it shall be well with

you." When she repeated her visit, the same things occurred, and she heard the most exquisite music.

The same sort of phenomena were witnessed by every body who went—the immoral were rebuked, the good encouraged. Some were bidden instantly depart, and were forced to go. The voices of several deceased persons of the family were also heard, and made revelations.

Once, the voice said, "Look up, and you shall see the sun and moon on the ceiling!" and immediately there appeared a beautiful representation of these planets in lively colours, viz., green, yellow, and orange. Moreover, these figures were permanent; but the father, who was a long time sceptical, insisted on white-washing them over; however, they still remained visible.

Amongst other things, the voice said, that though the child appeared to suffer, that she did not; that she did not know where her body was, and that her own spirit had left it, and another had entered; and that her body was made a speaking-trumpet. The voice told the family and visitors many things of their distant friends, which proved true.

The girl twice saw a divine form standing by her bed-side who spoke to her, and Joseph

Ragg, one of the persons who had been invited
by the voice to go, saw a beautiful and
heavenly figure come to his bedside about
eleven o'clock at night, on the 17th January. It
was in male attire, surrounded by a radiance ;
it came a second time on the same night.   On
each occasion it opened his curtains and looked
at him benignantly, remaining about a quarter
of an hour.   When it went away, the curtains
fell back into their former position.   One day
whilst in the sick child's room, Margaret
Watson saw a lamb, which passed through the
door and entered a place where the father,
John Jobson, was; but he did not see it.

One of the most remarkable features in this
case, is the beautiful music which was heard
by all parties, as well as the family, including
the unbelieving father, and, indeed, it seems to
have been, in a great degree, this that converted him at last.   This music was heard repeatedly during a space of sixteen weeks;
sometimes it was like an organ, but more
beautiful ; at others, there was singing of holy
songs, *in parts*, and the words distinctly
heard.   The sudden appearance of water in
the room, too, was most unaccountable; for
they felt it, and it was really water. When the
voice desired that water should be sprinkled,

it immediately appeared as if sprinkled. At
another time a sign being promised to the
sceptical father, water would suddenly appear
on the floor; this happened "not once, but
twenty times."

During the whole course of this affair, the
voices told them that there was a miracle to be
wrought on this child; and accordingly on the
22nd of June, when she was as ill as ever, and
they were only praying for her death, at five
o'clock the voice ordered that her clothes
should be laid out, and that everybody should
leave the room, except the infant which was
two years and a half old. They obeyed; and
having been outside the door a quarter of an
hour, the voice cried "Come in !" and when
they entered, they saw the girl completely
dressed and quite well, sitting in a chair with
the infant on her knee, and she had not had an
hour's illness from that time till the report was
published, which was on the 30th of January,
1841.

Now, it is very easy to laugh at all this,
and assert that these things never happened,
because they are absurd and impossible; but
whilst honest, well-meaning, and intelligent
people, who were on the spot, assert that they
did, I confess I find myself constrained to

believe them, however much I find in the
case which is discrepent with my notions. It
was not an affair of a day or an hour; there
was ample time for observation, for the
phenomena continued from the 9th of February to the 22nd of June; and the determined
unbelief of the father, with regard to the possibility of spiritual appearances, insomuch,
that he ultimately expressed great regret for
the harshness he had used—is a tolerable
security against imposition. Moreover, they
pertinaciously refused to receive any money
or assistance whatever, and were more likely
to suffer in public opinion than otherwise by
the avowal of these circumstances.

Dr. Reid Clanny, who publishes the report,
with the attestations of the witnesses, is a
physician of many years experience, and is
also, I believe, the inventor of the improved
Davy Lamp; and he declares his entire conviction of the facts, assuring his readers, that
" many persons holding high rank in the
Established Church, ministers of other denominations, as well as many lay-members of
society, highly respected for learning and
piety, are equally satisfied." When he first
saw the child lying on her back, apparently
insensible, with her eyes suffused with florid

blood, he felt assured that she had a disease of the brain; and he was not in the least disposed to believe in the mysterious part of the affair, till subsequent investigation compelled him to do so; and that his belief is of a very decided character we may feel assured, when he is content to submit to all the obloquy he must incur by avowing it.

He adds, that since the girl has been quite well, both her family and that of Joseph Ragg, have frequently heard the same heavenly music, as they did during her illness; and a Mr. Torbock, a surgeon, who expresses himself satisfied of the truth of the above particulars, also mentions another case, in which he, as well as a dying person he was attending, heard divine music just before the dissolution.

Of this last phenomenon, namely, sounds as of heavenly music, being heard when a death was occurring, I have met with numerous instances.

From investigation of the above case, Dr. Clanny has arrived at the conviction that the spiritual world do occasionally identify themselves with our affairs; and Dr. Drury asserts that besides this instance he has met with another circumstance which has left him firmly convinced that we live in a world of spirits, and that he has been in the presence of an unearthly being, who had "passed that bourne, from which (it is said) no traveller returns." *

But the most extraordinary case I have yet met with is the following ; because it is one which cannot by any possibility be attributed to disease or illusion. It is furnished to me from the most undoubted authority, and I give it as I received it, with the omission of the names. I have indeed, in this instance, thought it right to change the initial, and substitute G. for the right one, the particulars being of a nature which demand the greatest delicacy, as regards the parties concerned :—

" Mrs. S. C. Hall, in early life, was intimately acquainted with a family, named G., one of whom, Richard G., a young officer in the army, was subject to a harrassing visitation of a kind that is usually regarded as super-natural. Mrs. H. once proposed to pay a visit to her particular friend, Catherine G., but was told that it would not be convenient exactly at that time, as Richard was on the point of com-ing home. She thought the inconvenience con-sisted in the want of a bed-room, and spoke of sleeping with Miss G., but found that the ob-jection really lay in the fact of Richard being

* Alluding, I conclude, to the affair at Willington.

" haunted," which rendered it impossible for anybody else to be comfortable in the same house with him. A few weeks after Richard's return, Mrs. H. heard of Mrs. G.'s being extremely ill; and found, on going to call, that it was owing to nothing but the distress the old lady suffered in consequence of the strange circumstance connected with her son. It appeared that Richard, wherever he was, at home, in camp, in lodgings, abroad or in his own country, was liable to be visited in his bedroom at night by certain extraordinary noises. Any light he kept in the room, was sure to be put out. Something went beating about the walls and his bed, making a great noise, and often shifting close to his face, but never becoming visible. If a cage-bird was in his room, it was certain to be found dead in the morning. If he kept a dog in the apartment, it would make away from him as soon as released, and never came near him again. His brother, even his mother, had slept in the room; but the visitation took place as usual. According to Miss G.'s report, she and other members of the family would listen at the bed-room door after Richard had gone to sleep, and would hear the noises commence; and they would then hear him sit up and express his vexation by a few military execrations. The young man, at length, was obliged by this pest to quit the army, and go upon half-pay. Under its influence he became a sort of Cain ; for wherever he lived, the annoyance was so great that he was quickly obliged to remove. Mrs. H. heard of him having ultimately gone to settle in Ireland, where, however, according to a brother whom she met about four years ago, the visitation which afflicted him in his early years, was in no degree abated."

This cannot be called a case of possession ; but seems to be ore of a rapport, which attaches this invisible tormentor to his victim.

CHAPTER VII.

MISCELLANEOUS PHENOMENA.

In a former chapter, I alluded to the forms
seen floating over graves, by Billing,
Pfeffel's amanuensis. By some persons, this
luminous form is seen only as a light, just as
occurs in many of the apparition cases I have
related. How far Baron Reichenbach is correct in his conclusion, that these figures are
merely the result of the chemical process going
on below, it is impossible for any one at present to say. The fact that these lights do not
always hover over the graves, but sometimes
move from them, militates against this

opinion, as I have before observed; and the insubstantial nature of the form which reconstructed itself after Pfeffel had passed his stick through it, proves nothing; since the same thing is asserted of all apparitions I meet with, let them be seen where they may, except in such very extraordinary cases as that of the Bride of Corinth, supposing that story to be true.

At the same time, although these cases are not made out to be chemical phenomena, neither are we entitled to class them under the head of what is commonly understood by the word *ghost;* whereby we comprehend a shadowy shape, informed by an intelligent spirit. But there are some cases, a few of which I will mention, that it seems extremely difficult to include under one category or the other.

The late Lieutenant-General Robertson of Lawers, who served during the whole of the American war, brought home with him, at its termination, a negro, who went by the name of Black Tom, and who continued in his service. The room appropriated to the use of this man in the General's town residence—I speak of Edinburgh—was on the ground floor; and he was heard frequently to complain that he

could not rest in it, for that every night the
figure of a headless lady, with a child in her
arms, rose out of the hearth and frightened
him dreadfully. Of course nobody believed
this story, and it was supposed to be the dream
of intoxication, as Tom was not remarkable
for sobriety ; but strange so say, when the old
mansion was pulled down to build Gillespie's
Hospital, which stands on its site, there was
found under the hearth-stone in that apartment,
a box containing the body of a female, from
which the head had been severed ; and beside
her lay the remains of an infant wrapt in a
pillow case, trimmed with lace. She appeared,
poor lady, to have been cut off in the " blos-
som of her sins ;" for she was dressed, and her
scissars were yet hanging by a ribbon to her
side, and her thimble was also in the box,
having, apparently, fallen from the shrivelled
finger.

Now, whether we are to consider this a
ghost, or a phenomenon of the same nature as
that seen by Billing, it is difficult to decide.
Somewhat similar is the following case, which
I have borrowed from a little work entitled
" Supernaturalism in New England." Not
only does this little extract prove that the
same phenomena, be they interpreted as they

may, exist in all parts of the world, but I think
it will be granted me, that although we have
not here the confirmation that time furnished
in the former instance, yet, it is difficult to
suppose that this unexcitable person should
have been the subject of so extraordinary a
spectral illusion.

"Whoever has seen Great Pond, in the
East parish of Haverhill, has seen one of the
very loveliest of the thousand little lakes or
ponds of New England. With its soft slopes
of greenest verdure—its white and sparkling
sand-rim—its southern hem of pine and maple,
mirrored, with spray and leaf, in the glassy
water—its gracefal hill-sentinels round about,
white with the orchard-bloom of spring, or
tasselled with the corn of autumn—its long
sweep of blue waters, broken here and there
by picturesque headlands—it would seem a
spot, of all others, where spirits of evil must
shrink, rebuked and abashed, from the presence of the beautiful. Yet here, too, has the
shadow of the supernatural fallen. A lady of
my acquaintance, a staid, unimaginative
church member, states that a few years ago,
she was standing in the angle formed by two
roads, one of which traverses the pond shore,
the other leading over the hill which rises ab-

ruptly from the water. It was a warm summer evening, just at sunset. She was startled by the appearance of a horse and cart of the kind used a century ago in New England, driving rapidly down the steep hill-side, and crossing the wall a few yards before her, without noise or displacing of a stone. The driver sat sternly erect, with a fierce countenance, grasping the reins tightly, and looking neither to the right nor the left. Behind the cart, and apparently lashed to it, was a woman of gigantic size, her countenace convulsed with a blended expression of rage and agony, writhing and struggling, like Laocoon in the folds of the serpent. Her head, neck, feet, and arms were naked; wild locks of grey hair streamed back from temples corrugated and darkened. The horrible cavalcade swept by across the street, and disappeared at the margin of the pond."

Many persons will have heard of the " Wild Troop of Rodenstein,"but few are aware of the curious amount of evidence there is in favour of the strange belief which prevails amongst the inhabitants of that region. The story goes, that the former possessor of the Castle of Rodenstein and Schnellert, were robbers and pirates, who committed, in conjunction, all manner of enormities; and that, to this day, the troop, with their horses and carriages and dogs are heard, every now and then, wildly rushing along the road betwixt the two castles.   This sounds like a fairy tale; yet so much was it believed, that up to the middle of the last century regular reports were made to the authorities in the neighbourhood, of the periods when the troop had passed.   Since that, the Landgericht or Court Leet, has been removed to Furth, and they trouble themselves no longer about the Rodenstein Troop; but a traveller named Wirth, who a few years ago undertook to examine into the affair, declares the people assert that the passage of the visionary cavalcade still continues; and they assured him that certain houses that he saw lying in ruins, were in that state, because, as they lay directly in the way of the troop, they were uninhabitable.   There is seldom anything seen, but the sound of carriage wheels, horses feet, smacking of whips, blowing of horns, and the voice of these fierce hunters of men urging them on, are the sounds by which they recognize that the troop is passing from one castle to the other; and at a spot which was formerly a blacksmith s, but is now a carpenter's, the invisible Lord of Rodenstein

2 E 5

still stops to have his horse shod. Mr. With copied several of the depositions out of the court records, and they are brought down to June 1764. This is certainly a strange story; but it is not much more so than that of the black man, which I know to be true.

During the seven years war in Germany, a drover lost his life in a drunken squabble on the high road.

For some time there was a sort of rude tomb-stone, with a cross on it, to mark the spot where his body was interred; but this has long fallen, and a mile-stone now fills its place. Nevertheless, it continues commonly asserted by the country people, and also by various travellers, that they have been deluded in that spot by seeing, as they imagine, herds of beasts, which on investigation prove to be merely visionary. Of course, many people look upon this as a superstition; but a very singular confirmation of the story occurred in the year 1826, when two gentlemen and two ladies were passing the spot in a post carriage. One of these was a clergyman, and none of them had ever heard of the phenomenon said to be attached to the place. They had been dis-cussing the prospects of the minister, who was on his way to a vicarage, to which he had just

been appointed, when they saw a large flock
of sheep, which stretched quite across the road,
and was accompanied by a shepherd and a long
haired black dog. As to meet cattle on that
road was nothing uncommon, and indeed they
had met several droves in the course of the day,
no remark was made at the moment, till suddenly each looked at the other and said,
"What is become of the sheep?" Quite perplexed
at their sudden disappearance, they called to the
postilion to stop, and all got out, in order
to mount a little elevation and look around,
but still unable to discover them, they now be-
thought themselves of asking the postilion
where they were; when, to their infinite surprise, they learnt that he had not seen them.
Upon this, they bade him quicken his pace,
that they might overtake a carriage that had
passed them shortly before, and enquire if that
party had seen the sheep ; but they had not.

Four years later, a postmaster, named J.,
was on the same road, driving a carriage, in
which were a clergyman and his wife, when he
saw a large flock of sheep near the same spot.
Seeing they were very fine wethers, and supposing them to have been bought at a sheep-
fair that was then taking place a few miles off,
J. drew up his reins and stopped his horses,

turning at the same time to the clergyman to
say, that he wanted to enquire the price of the
sheep, as he intended going next day to the
fair himself. Whilst the minister was asking
him what sheep he meant, J. got down and
found himself in the midst of the animals, the
size and beauty of which astonished him. They
passed him at an unusual rate, whilst he made
his way through them to find the shepherd,
when on getting to the end of the flock, they
suddenly disappeared. He then first learnt
that his fellow travellers had not seen them
at all.

Now, if such cases as these are not pure
illusions, which I confess I find it difficult
to believe, we must suppose that the animals
and all the extraneous circumstances are produced by the mgical will of the spirit, either
acting on the constructive imagination of the
seers, or else actually constructing the etherial
forms out of the elements at its command;
just as we have supposed an apparition able
to present himself with whatever dress or
appliances he conceives; or else we must conclude, these forms to have some relation to the
mystery called PALINGNESIA which I have
previously alluded to; although the motion
and change of place renders it difficult to bring

them under this category. As for the animals, although the drover was slain, they were not; and therefore even granting them to have souls, we cannot look upon them as the apparitions of the flock. Neither can we consider the numerous instances of armies seen in the air to be apparitions; and yet these phenomena are so well established, that they have been accounted for by supposing them to be atmospherical reflections of armies elsewhere, in actual motion. But how are we to account for the visionary troops which are not seen in the air, but on the very ground on which the seers themselves stand? which was the case especially with those seen in Havarah Park, near Ripley, in the year 1812. These soldiers wore a white uniform, and in the centre was a personage in a scarlet one.

After performing several evolutions. the body began to march in perfect order to the summit of a hill, passing the spectators at the distance of about one hundred yards. They amounted to several hundreds, and marched in a column, four deep, across about thirty acres; and no sooner were they passed, than another body, far more numerous, but dressed in dark clothes arose and marched after them, without any apparent hostility. Both parties having reached the top of the hill, and there formed what the spectators called an L, they disappeared down the other side, and were seen no more; but at that moment, a volume of smoke arose like the discharge of a park of artillery, which was so thick that the men could not, for two or three minutes, discover their own cattle. They then hurried home to relate what they had seen, and the impression made on them is described as so great, that they could never allude to the subject without emotion.

One of them was a farmer of the name of Jackson, aged forty-five; the other was a lad of fifteen, called Turner; and they were at the time herding cattle in the park. The scene seems to have lasted nearly a quarter of an hour, during which time they were quite in possession of themselves, and able to make remarks to each other on what they saw. They were both men of excellent character and unimpeachable veracity, insomuch that nobody who knew them doubted that they actually saw what they described, or, at all events, believed that they did. It is to be observed also, that the ground is not swampy, nor subject to any exhalations.

About the year 1750, a visionary army of the

same description was seen in the neighbourhood of Inverness, by a respectable farmer, of
Glenary, and his son. The number of troops
was very great, and they had not the slightest
doubt that they were otherwise than sub-
stantial forms of flesh and blood. They counted
at least sixteen pairs of columns, and had
abundance of time to observe every particular.
The front ranks marched seven abreast, and
were accompanied by a good many women and
children, who were carrying tin cans and other
implements of cookery. The men were clothed
in red, and their arms shone brightly in the
sun. In the midst them was an animal, a
deer or a horse, they could not distinguish
which, that they were driving furiously forward with their bayonets. The younger of
the two men observed to the other, that every
now and then, the rear ranks were obliged to
run to overtake the van; and the elder one,
who had been a soldier, remarked that that
was always the case, and recommended him, if
he ever served, to try and march in the front.
There was only one mounted officer; he rode
a grey dragoon horse, and wore a gold-laced
hat, and blue Hussar cloak, with wide open
sleeves lined with red. The two spectators
observed him so particularly, that they said

afterwards, they should recognize him anywhere. They were, however, afraid of being
ill-treated, or forced to go along with the troops,
whom they concluded had come from Ireland,
and landed at Kyntyre; and whilst they were
climbing over a dyke to get out of their way,
the whole thing vanished.

Some years since, a phenomenon of the
same sort was observed at Paderborn, in West-
phalia, and seen by at least thirty persons, as
well as by horses and dogs, as was discovered
by the demeanour of these animals, In October, 1836, on the very same spot, there was
a review of twenty thousand men ; and the
people then concluded, that the former vision
was a *second sight.*

A similar circumstance occurred in Stockton
Forest, some years ago ; and there are many
recorded elsewhere ; one especially, in the
year 1686, near Lanark, where, for several
afternoons, in the months of June and July,
there were seen, by numerous spectators, companies of men in arms, marching in order by
the banks of the Clyde, and other companies
meeting them, &c. &c. ; added to which, there
were showers of bonnets, hats, guns, swords,
&c., which the seers described with the greatest
exactness. All who were present could not

see these things, and Walker relates, that one gentleman, particularly, was turning the thing into ridicule, calling the seers "Damned witches and warlocks, with the second sight!" boasting that "The devil a thing he could see!" when he suddenly exclaimed, with fear and trembling, that he now saw it all; and entreated those who did not see, to say nothing—a change that may be easily accounted for, be the phenomena of what nature it may, by supposing him to have touched one of the seers, when the faculty would be communicated like a shock of electricity.

With regard to the palingnesia, it would be necessary to establish that these objects had previously existed, and that, as Oetinger says, the earthly husk having fallen off, " the volatile essence had ascended perfect in form, but void of substance."

The notion supported by Baron Reichenbach that the lights seen in churchyards and over graves are the result of a process going on below, is by no means new; for Gaffarillus suggested the same opinion in 1650; only he speaks of the appearances over graves and in churchyards as shadows, *ombres*, as they appeared to Billing; and he mentions, casually, as a thing frequently observed, that the same visionary forms are remarked on ground where battles have been fought, which he thinks arise out of a process betwixt the earth and the sun. When a limb has been cut off, some somnambules still discern the form of the member as if actually attached.

But this magical process is said to be not only the work of the elements, but also possible to man; and that as the forms of plants can be preserved after the substance is destroyed, so can that of man, be either preserved or reproduced from the elements of his body. In the reign of Louis XIV. three alchemists having distilled some earth, taken from the Cemetery of the Innocents, in Paris, were forced to desist, by seeing the forms of men appearing in their vials, instead of the philosopher's stone, which they were seeking, and a physician, who, after dissecting a body, and pulverising the cranium, which was then an article, admitted into the *materia medica*, had left the powder on the table of his laboratory, in charge of his assistant, the latter, who slept in an adjoining room, was awakened in the night, by hearing a noise, which, after some search, he ultimately traced to the powder; in the midst of which he beheld, gradually constructing itself, a human form. First appeared the head, with two open eyes, then the arms and hands, and by degrees, the rest of the person, which subsequently assumed the clothes it had worn when alive. The man was of course frightened out of his wits ; the rather, as the apparition planted itself before the door, and would not let him go away, till it had made its own exit, which it speedily did. Similar results have been said to arise from experiments performed on blood. I confess I should be disposed to consider these apparitions, if ever they appeared, cases of genuine ghosts, brought into rapport by the operations, rather than forms residing in the bones or blood. At all events, these things are very hard to believe; but seeing we were not there, I do not think we have any right to say they did not happen ; or at least that some phenomena did not occur, that were open to this interpretation.

It is highly probable that the seeing of those visionary armies and similar prodigies is a sort of second sight ; but having admitted this, we are very little nearer an explanation. Granting that, as in the above experiments, the essence of things may retain the forms of the substance, this does not explain the seeing that which has not yet taken place, or which is taking place at so great a distance, that neither Oetinger's essence nor the superficial films of Lucretius can remove the difficulty.

It is the fashion to say, that second sight was a mere superstition of the Highlanders, and that no such thing is ever heard of now; but those who talk in this way know very little of the matter. No doubt, if they set out to look for seers, they may not find them; such phenomena, though known in all countries, and in all ages, are *comparatively* rare, as well as uncertain and capricious; and not to be exercised at will; but I know of too many instances of the existence of this faculty in families, as well as of isolated cases occurring to individuals above all suspicion, to entertain the smallest doubt of its reality. But the difficulty of furnishing evidence is considerable; because, when the seers are of the humbler classes, they are called impostors and not believed; and when they are of the higher, they do not make the subject a matter of conversation, nor choose to expose themselves to the ridicule of the foolish; and consequently the thing is not known beyond their own immediate friends. When the young Duke of Orleans was killed, a lady residing here, saw the accident, and described it to her husband at the time it was occurring in France. She had frequently seen the Duke, when on the continent.

Captain N. went to stay two days at the house of Lady T. After dinner, however, he announced that he was under the necessity of going away that night, nor could he be induced to remain. On being much pressed for an explanation, he confided to some of the party, that during the dinner he had seen a female figure with her throat cut, standing behind Lady T.'s chair. Of course, it was thought an illusion, but Lady T. was not told of it, lest she should be alarmed. That night the household was called up for the purpose of summoning a surgeon—Lady T. had cut her own throat.

Mr. C., who, though a Scotchman, was an entire sceptic with regard to the second sight, was told by a seer whom he had been jeering on the subject, that within a month, he (Mr. C.) would be a pall-bearer at a funeral, that he would go by a certain road, but that before they had crossed the brook, a man in a drab coat would come down the hill and take the pall from him. The funeral occurred, Mr. C. was a bearer, and they went by the road described; but he firmly resolved that he would disappoint the seer by keeping the pall whilst they crossed the brook ; but shortly before they reached it, the postman overtook them, with letters, which in that part of the country arrived but twice a week, and Mr. C., who was engaged in some speculations of importance, turned to receive them ; at which moment the pall was taken from him, and on looking round, he saw it was by a man in a drab coat.

A medical friend of mine, who practised some time at Deptford, was once sent for to a girl who had been taken suddenly ill. He found her with inflammation of the brain, and the only account the mother could give of it was, that shortly before, she had ran into the room, crying, " Oh mother, I have seen Uncle John drowned in his boat under the fifth arch of Rochester Bridge ! " The girl died a few hours afterwards ; and on the following night, the uncle's boat ran foul of the bridge, and he was drowned, exactly as she had foretold.

Mrs. A., an English lady, and the wife of a clergyman, relates that, previous to her marriage, she with her father and mother being at the seaside, had arranged to make a few days' excursion to some races that were about to take place; and that the night before they started, the father having been left alone, whilst the ladies were engaged in their prepa-

rations, they found him, on descending to the
drawing-room, in a state of considerable agitation ; which, he said, had arisen from his
having seen a dreadful face at one corner of the
room. He described it as a bruised, battered,
crushed, discoloured face, with the two eyes
protruding frightfully from their sockets ; but
the features were too disfigured to ascertain if
it were the face of any one he knew. On the
following day, on their way to the races, an
accident occurred ; and he was brought home
with his own face exactly in the condition he
had described. He had never exhibited any
other instance of this extraordinary faculty, and
the impression made by the circumstance lasted
the remainder of his life, which was unhappily
shortened by the injuries he had received.

The late Mrs. V., a lady of fortune and
family, who resides near Loch Lomond, possessed this faculty in an extraordinary degree ;
and displayed it on many remarkable occasions. When her brother was shipwrecked
in the Channel, she was heard to exclaim,
" Thank God, he is saved!" and described the
scene, with all its circumstances.

Colonel David Stewart, a determined disbeliever in what he calls *the supernatural*, in
his book on the Highlanders, relates the

following fact as one so remarkable, that
" credulous minds " may be excused for believing it to have been prophetic. He says
that late in an autumnal evening of the year
1773, the son of a neighbour came to his
father's house, and soon after his arrival
enquired for a little boy of the family, then
about three years old. He was shown up to
the nursery and found the nurse putting a
pair of new shoes on the child, which she
complained did not fit. " Never mind," said the
young man, " they will fit him before he wants
them," a prediction which not only offended
the nurse, but seemed at the moment absurd,
since the child was apparently in perfect
health. When he joined the party in the
drawing room, he being much jeered upon
this new gift of second sight, he explained,
that the impression he had received originated
in his having just seen a funeral passing the
wooden bridge which crossed a stream at a
short distance from the house. He first observed a crowd of people, and on coming
nearer, he saw a person carrying a small
coffin, followed by about twenty gentlemen,
all of his acquaintance, his own father and a
Mr. Stewart being amongst the number.
He did not attempt to join the procession,

which he saw turn off into the churchyard ; but knowing his own father could not be actually there, and that Mr. and Mrs. Stewart were then at Blair, he felt a conviction that the phenomenon portended the death of the child ; a persuasion which was verified by its suddenly expiring on the following night, and Colonel Stewart adds, that the circumstances and attendants at the funeral were precisely such as the young man had described. He mentions also that this gentleman was not a seer ; that he was a man of education and general knowledge, and that this was the first and only vision of the sort he ever had.

I know of a young lady, who has three times seen funerals in this way.

The old persuasion, that fasting was a means of developing the spirit of prophesy, is undoubtedly well founded, and the annals of medicine furnish numerous facts which establish it. A man condemned to death at Viterbo having abstained from food in the hope of escaping execution, became so *clairvoyant*, that he could tell what was doing in any part of the prison ; the expression used in the report is, that he *saw through the walls*, this, however, could not be with his natural organs of sight.

It is worthy of observation, that idiots often

possess some gleams of this faculty of second sight or presentiment; and it is probably on this account that they are in some countries held sacred. Presentiment, which I think may very probably be merely the vague and imperfect recollection of what we *knew* in our sleep, is often observed in drunken people.

In the great plague at Basle, which occurred towards the end of the sixteenth century, almost everybody who died, called out in their last moments, the name of the person that was to follow them next.

Not long ago a servant girl on the estate of D., of S., saw with amazement five figures ascending a perpendicular cliff, quite inaccessible to human feet; one was a boy wearing a cap with red binding. She watched them with great curiosity till they reached the top, where they all stretched themselves on the earth, with countenances expressive of great dejection. Whilst she was looking at them they disappeared, and she immediately related her vision. Shortly afterwards, a foreign ship in distress, was seen to put off a boat with four men and a boy; the boat was dashed to pieces in the surf, and the five bodies, exactly answering the description she had given, were thrown on shore, at the foot of the cliff, which they had perhaps climbed in the spirit!

How well what we call *clairvoyance* was known, though how little understood, at the period of the witch persecution, is proved by what Dr. Henry More says, in his " Antidote against Atheism" :—

" We will now pass to those supernatural effects which are observed in them that are bewitched or possessed; and such as foretelling things to come, telling what such and such persons speak or do, as exactly as if they were by them, when the party possessed is at one end of the town, and sitting in a house within doors, and those parties that act and confer together are without, at the other end of the town ; to be able to see some, and not others ; to play at cards with one certain person, and not to discern anybody else at the table beside him ; to act and talk, and go up and down, and tell what will become of things, and what happens in those fits of possession; and then so soon as the possessed or bewitched party is out of them, to remember nothing at all, but to enquire concerning the welfare of those whose faces they seemed to look upon but just before, when they were in their fits."

A state which he believes to arise from the devil's having taken possession of the body of the magnetic person, which is precisely the

theory supported by many fanatical persons in
our own day. Dr. More was not a fanatic ;
but these phenomena, though very well
understood by the ancient philosophers, as well
as by Paracelsus, Van Helmont, Cornelius
Agrippa, Jacob Behmen, a Scotch physician
called Maxwell, who published on the subject
in the seventeenth century, and many others,
were still, when observed, looked upon as the
effects of diabolical influence by mankind in
general.

When Monsieur Six Deniers, the artist, was
drowned in the Seine, in 1846, after his body
had been vainly sought, a somnambule was
applied to, in whose hands they placed a port-
folio belonging to him, and being asked where
the owner was, she evinced great terror, held up
her dress, as if walking in the water, and said
that he was between two boats, under the
Pont des Arts, with nothing on but a flannel
waistcoat ; and there he was found.

A friend of mine knows a lady, who, one
morning, early, being in a natural state of
clairvoyance, without magnetism, saw the
porter of the house where her son lodged,
ascend to his room with a carving-knife, go to
his bed where he lay asleep, lean over him,
then open a chest, take out a fifty-pound note,

and retire. On the following day, she went to
her son and asked him if he had any money
in the house; he said, " Yes, he had fifty
pounds ;" whereupon, she bade him seek it;
but it was gone. They stopped payment of
the note ; but did not prosecute, thinking the
evidence insufficient. Subsequently, the porter
being taken up for other crimes, the note was
found crumpled up at the bottom of an old
purse belonging to him.

Dr. Ennemoser says, that there is no doubt
of the ancient Sybyls having been *clairvoyante*
women, and that it is impossible so much
value could have been attached to their books,
had not their revelations been verified.

A maid-servant, residing in a family in
Northumberland, one day, last winter, was
heard to utter a violent scream immediately
after she had left the kitchen. On following her
to enquire what had happened, she said that
she had just seen her father in his night
clothes, with a most horrible countenance, and
she was sure something dreadful had happened
to him. Two days afterwards, there arrived
a letter, saying, he had been seized with
delirium tremens, and was at the point of
death ; which accordingly ensued.

There are innumerable cases of this sort

recorded in various collections; not to mention the much more numerous ones that meet
with no recorder; and I could myself mention
many more, but these will suffice—one, however, I will not omit, for though historical it
is not generally known. A year before the
rebellion broke out, in consequence of which
Lord Kilmarnock lost his head, the family
were one day startled by a violent scream, and
on rushing out to enquire what had occurred,
they found the servants all assembled in
amazement, with the exception of one maid, who
they said had gone up to the garrets to hang
some linen on the lines to dry. On ascending
thither, they found the girl on the floor, in a
state of insensibility; and they had no sooner
revived her, than on seeing Lord Kilmarnock
bending over her, she screamed and fainted
again. When ultimately recovered, she told
them that whilst hanging up her linen and
singing, the door had burst open and his
lordships bloody head had rolled in. I think
it came twice. This event was so well known
at the time, that on the first rumours of the
rebellion, Lord Saltoun said, "Kilmarnock
will lose his head." It was answered "that
Kilmarnock had not joined the rebels." "He
will, and will be beheaded," returned Lord S.

Now, in these cases we are almost compelled to believe that the phenomenon is purely subjective, and that there is no veritable outstanding object seen; yet, when we have taken refuge in this hypothesis, the difficulty remains as great as ever; and is to me much more incomprehensible than ghost-seeing, because in the latter we suppose an external agency acting in some way or other on the seer.

I have already mentioned that Oberlin, the good pastor of Ban de la Roche, himself a ghost-seer, asserted that everything earthly had its counterpart, or antitype, in the other world, not only organized, but unorganized matter. If so, do we sometimes see these antitypes?

Dr. Ennemoser, in treating of second sight —which, by the way, is quite as well known in Germany, and especially in Denmark, as in the Highlands of Scotland—says, that as in natural somnambulism, there is a partial internal vigilance, so does the seer fall, whilst awake, into a dream-state. He suddenly becomes motionless and stiff: his eyes are open, and his senses are, whilst the vision lasts, unperceptive of all external objects; the vision may be communicated by the touch, and sometimes persons at a distance from each other, but connected by blood or sympathy have the vision simultaneously. He remarks, also, that, as we have seen in the above-mentioned case of Mr. C., any attempt to frustrate the fulfilment of the vision never succeeds, inasmuch as the attempt appears to be taken into the account.

The seeing in glass and in crystals, is equally inexplicable; as is the magical seeing of the Egyptians. Every now and then, we hear it said that this last is discovered to be an imposition, because some traveller has either actually fallen into the hands of an impostor—and there are impostors in all trades; or because the phenomenon was imperfectly exhibited; a circumstance which, as in the exhibitions of clairvoyants and somnambulists, where all the conditions are not under command, or even recognized, must necessarily happen. But not to mention the accounts published by Mr. Lane and Lord Prudhoe, whoever has read that of Monsieur Léon Laborde, must be satisfied that the thing is an indisputable fact. It is, in short, only another form of the seeing in crystals, which has been known in all ages, and of which many modern instances have occurred amongst somnambulic patients.

We see by the 44th chapter of Genesis that it

was by his cup that Joseph prophesied: " Is
not this it in which my lord drinketh, and
whereby indeed he divineth?"   But, as Dr.
Passavent observes, and as we shall presently
see in the anecdote of the boy and the gipsy, the
virtue does not lie in the glass nor in the
water, but in the seer himself, who may
possess a more or less developed faculty.   The
external objects and ceremonies being only the
means of concentrating the attention and inten-
sifying the power.

Monsieur Léon Laborde witnessed the exhibition, at Cairo, before Lord P.'s visit ; the
exhibitor, named Achmed, appeared to him a
respectable man, who spoke simply of his
science, and had nothing of the charlatan
about him.   The first child employed, was a
boy of eleven years old, the son of a European ;
and Achmed having traced some figures on the
palm of his hand, and poured ink over them,
bade him look for the reflection of his own
face.   The child said he saw it; the magician
then burnt some powders in a brazier, and
bade him tell him when he saw a soldier
sweeping a place; and whilst the fumes from
the brazier diffused themselves, he pronounced
a sort of litany.   Presently, the child threw
back his head, and screaming with terror,

sobbed out, whilst bathed in tears, that he had seen a dreadful face. Fearing the boy might be injured, Monsieur Laborde now called up a little Arab servant, who had never seen or heard of the magician. He was gay and laughing, and not at all frightened; and the ceremony being repeated, he said he saw the soldier sweeping in the front of a tent. He was then desired to bid the soldier bring Shakspeare, Colonel Cradock, and several other persons; and he described every person and thing so exactly, as to be entirely satisfactory. During the operations, the boy looked as if intoxicated; with his eyes fixed and the perspiration dripping from his brow. Achmed disenchanted him by placing his thumbs on his eyes; he gradually recovered, and gaily related all he had seen, which he perfectly remembered.

Now this is merely another form of what the Laplanders, the African magicians, and the Schaamans of Siberia, do by taking narcotics and turning round till they fall down in a state of insensibility, in which condition they are clear-seers, and besides vaticinating, describe scenes, places, and persons they have never seen. In Barbary they anoint their hands with a black ointment, and then holding them up in the sun, they see whatever they desire, like the Egyptians.

Lady S. possesses somewhat of a singular faculty, naturally. By walking rapidly round a room, several times, till a certain degree of vertigo is produced, she will name to you any person you have privately thought of, or agreed upon with others. Her phrase is, " I *see* so and so."

Monsieur Laborde purchased the secret of Achmed, who said, he had learnt it from two celebrated Scheicks of his own country, which was Algiers. Monsieur L. found it connected both with physics and magnetism, and he practised it himself afterwards with perfect success, and he affirms, positively, that under the influence of a particular organization and certain ceremonies, amongst which he cannot distinguish which are indispensable and which are not, that a child without fraud or collusion, can see as through a window or peep-hole, people moving, who appear and disappear at their command, and with whom they hold communication—and they remember everything after the operation. He says " I narrate, but explain nothing ; I produced those effects, but cannot comprehend them ; I only affirm in the most positive manner, that what I relate is true. I performed the experiment in various places, with various subjects, before numerous witnesses, in my own room or other rooms, in the open air, and even in a boat on the Nile. The exactitude and detailed descriptions of persons, places, and scenes, could by no possibility be feigned."

Moreover, Baron Dupotet has very lately succeeded in obtaining these phenomena in Paris, from persons, not somnambulic, selected from his audience; the chief difference being, that they did not recollect what they had seen when the crisis was over.

Cagliostro, though a charlatan, was possessed of this secret; and it was his great success in it, that chiefly sustained his reputation; the spectators, convinced he could make children see distant places and persons in glass, were persuaded he could do other things, which appeared to them no more mysterious. Dr. Dee was perfectly honest, with regard to his mirror in which he could *see*, by concentrating his mind on it; but as he could not remember what he saw, he employed Kelly to *see* for him, whilst he himself wrote down the revelations; and Kelly was a rogue, and deceived and ruined him.

A friend of Pfeffel's knew a boy, apprenticed to an apothecary, at Schoppenweyer, who, having been observed to amuse himself by looking into vials filled with water, was asked what he saw; when it was discovered that he possessed this faculty of *seeing* in glass, which was afterwards very frequently exhibited for the satisfaction of the curious. Pfeffel also mentions another boy, who had this faculty, who went about the country with a small mirror, answering questions, recovering stolen goods, and so forth. He said that he one day fell in with some gipsies, one of whom was sitting apart, and staring into this glass. The boy, from curiosity, looked over his shoulder, and exclaimed that he saw "a fine man, who was moving about;" whereupon the gipsy, having interrogated him, gave him the glass; "For," said he, "I have been staring in it long enough, and can see nothing but my own face."

It is almost unnecessary to observe, that the sacred books of the Jews, and of the Indians, testify to their acquaintanc e with this mode of divination, as well as many others.

Many persons will have heard or read an account of Mr. Canning and Mr. Huskisson having seen, whilst in Paris, the visionary representation of their own deaths in water, as exhibited to them by a Russian or Polish lady, there; as I do not, however, know what authority there is for this story, I will not insist on it here. But St. Simon relates a very curious circumstance of this nature, which occurred at Paris, and was related to him by the Duke of Orleans, afterwards Regent. The latter said that he had sent on the preceding evening for a man, then in Paris, who pretended to exhibit whatever was desired in a glass of water. He came, and a child of seven years old, belonging to the house, being called up, they bade her tell what she saw doing in certain places. She did; and, as they sent to these places, and found her report correct, they bid her next describe under what circumstances the King would die; without, however, asking when the death would take place.

The child knew none of the Court, and had never been at Versailles; yet she described everything exactly—the room, bed, furniture, and the King himself; Madame de Maintenon, Fagon, the physician, the princes and princesses—everybody, in short, including a child wearing an order, in the arms of a lady, whom she recognised as having seen, this was Madame de Ventadour.

It was remarkable, that she omitted the Dukes de Bourgogne and Berry, and Monseigneur, and also the Duchess de Bourgogne. Orleans insisted they must be there, describing them; but she always said "*No*." These persons were then all well; but they died before the King. She also saw the children of the Prince and Princess of Conti, but not themselves; which was correct, as they also died shortly after this occurrence.

Orleans then wished to see his own destiny; and the man said, if he would not be frightened he could show it him, as if painted on the wall; and after fifteen minutes of conjuration, the Duke appeared, of the natural size, dressed as usual, but with a *couronne fermée* or closed crown on his head, which they could not comprehend, as it was not that of any country they knew of. It covered his head, had only four circles, and nothing at the top. They had never seen such an one. When he became Regent, they understood that that was the interpretation of the prediction.

In connection with this subject, the aversion

to glass frequently manifested by dogs, is well worthy of observation.

When facts of this kind are found to be recorded, or believed in, in all parts of the world, from the beginning of it up to the present time, it is surely vain for the so called *savants* to deny them ; and as Cicero justly says, in describing the different kinds of magic, " What we have to do with, is the facts, since of the cause we know little. Neither," he adds, " are we to repudiate these phenomena, because we sometimes find them imperfect, or even false, any more than we are to distrust that the human eye sees, although some do this very imperfectly, or not at all."

We are part spirit and part matter ; by the former we are allied to the spiritual world and to the absolute spirit; and as nobody doubts that the latter can work magically, that is, by the mere act of will—for by the mere act of will all things were created, and by its constant exertion all things are sustained—why should we be astonished that we, who partake of the divine nature, and were created after God's own image, should also, within certain limits, partake of this magical power? That this power has been frequently abused, is the fault of those, who, being capable, refuse to investigate, and deny the existence of these and similar phenomena; and by thus casting them out of the region of legitimate science, leave them to become the prey of the ignorant and designing.

Dr. Ennemoser, in his very learned work on magic, shows us that all the phenomena of magnetism and somnambulism, and all the various kinds of divination, have been known and practised in every country under the sun; and have been intimately connected with, and, indeed, may be traced up to the fountain-head of every religion.

What are the limits of these powers possessed by us whilst in the flesh, how far they may be developed, and whether, at the extreme verge of what we can effect, we begin to be aided by God or by spirits of other spheres of existence bordering on ours, we know not; but, with respect to the morality of these practices, it suffices, that what is good in act or intention must come of good; and what is evil in act or intention, must come of evil; which is true now, as it was in the time of Moses and the prophets, when miracles and magic were used for purposes holy and unholy, and were to be judged accordingly. God works by natural laws, of which we yet know very little, and, in some departments of his kingdom, nothing; and what appears to us supernatural, only appears so from our ignorance; and whatever faculties or powers he has endowed us with, it must have been designed we should exercise and cultivate for the benefit and advancement of our race; nor can I for one moment suppose, that, though like everything else liable to abuse, the legitimate exercise of these powers, if we knew their range, would be useless, much less pernicious or sinful.

Of the magical power of *will*, as I have said before, we know nothing; and it does not belong to a purely rationalistic age to acknowledge what it cannot understand. In all countries men have arisen, here and there, who *have* known it, and some traces of it have survived both in language and in popular superstitions. "If ye have faith as a grain of mustard seed, ye shall say unto this mountain, Remove hence, and it shall remove; and nothing shall be impossible to you. Howbeit this kind goeth not out but by prayer and fasting." And, *veuillez et croyez*, will and believe, was the solution Puysegur gave of his magical cures; and no doubt the explanation of those affected by royal hands, is to be found in the fact, that they believed in *themselves*, and having *faith*, they could exercise *will*. But, with the belief in the divine right of kings, the faith and the power would naturally expire together.

With respect to what Christ says in the above-quoted passage of *fasting*, numerous instances are extant, proving that clear-seeing and other magical or spiritual powers are sometimes developed by it.

Wilhelm Krause, a doctor of philosophy and a lecturer at Jena, who died during the prevalence of the cholera, cultivated these powers and preached them. I have not been able to obtain his works, they being suppressed as far as is practicable by the Prussian Government. Krause could leave his body, and, to all appearance, die whenever he pleased. One of his disciples, yet living, the Count Von Eberstein possesses the same faculty.

Many writers of the sixteenth century were well acquainted with the power of will, and to this was attributed the good or evil influence of blessings and curses. They believed it to be of great effect in curing diseases, and that by it alone life might be extinguished. That, *subjectively*, life may be extinguished, we have seen by the cases of Colonel Townshend, the Dervish that was buried, Hermotinus and others: for doubtless the power that could perform so much, could under an adequate motive, have performed more: and since all things in nature, spiritual and material, are connected, and that there is an unceasing interaction betwixt them, we being members of one great whole, only individualised by our organisms, it is possible to conceive that the power which can be exerted on our own organism might be extended to others: and since we cannot conceive man to be an isolated being—the only intelligence besides God—none above us and none below—but must, on the contrary, believe that there are numerous grades of intelligences, it seems to follow, of course, that we must stand in some kind of relation to them, more or less intimate; nor is it at all surprising that with some individuals this relation should be more intimate than with others. Finally, we are not entitled to deny the existence of this magical or spiritual power, either as exerted by incorporated or unincorporated spirits, because we do not comprehend how it can be exerted; since in spite of all the words that have been expended on the subject, we are equally ignorant of the mode in which our own will acts upon our own muscles. We know the fact, but not the mode of it.

CHAPTER VIII.

CONCLUSION.

OF the power of the mind over matter, we have a remarkable example in the numerous well-authenticated instances of the *Stigmata*. As in most cases this phenomenon has been connected with a state of religious exaltation, and has been appropriated by the Roman Church as a miracle, the fact has been in this country pretty generally discredited; but without reason; Ennemoser, Passavent, Schubert, and other eminent German physiologists, assure us that not only is the fact perfectly established, as regards many of the so-called saints, but also that there have been indubitably modern instances, as in the case of the Extaticas of the Tyrol, Catherine Emmerich, commonly called the Nun of Dulmen, Maria Morl, and Domenica Lazzari, who have all exhibited the stigmata.

Catherine Emmerich, the most remarkable of the three, began very early to have visions, and to display unusual endowments. She was very pious: could distinguish the qualities of plants, reveal secrets or distant circumstances, and knew people's thoughts; but was however, extremely sickly, and exhibited a variety of extraordinary and distressing symptoms, which terminated in her death. The wounds of the crown of thorns round her head, and those of the nails in her hands and feet, were as perfect as if painted by an artist, and they bled regularly on Fridays. There was also a double cross on her breast. When the blood was wiped away, the marks looked like the puncture of flies. She seldom took any nourishment but water, and having been but a poor cow-keeper, she discoursed, when in the extatic state, as if inspired.

I am well aware that on reading this, many persons who never saw her will say it was all imposture. It is very easy to say this; but it is as absurd as presumptuous to pronounce on what they have had no opportunity of observing. I never saw these women either; but I find myself much more disposed to accept the evidence of those who did, than of those who only " do not believe, because they do not believe."

Neither Catherine Emmerich nor the others, made their sufferings a source of profit, nor had they any desire to be exhibited; but quite the contrary. She could see in the dark as well as the light, and frequently worked all night at making clothes for the poor, without lamp or candle.

There have been instances of magnetic patients being stigmatised in this manner, Madam B. Von N. dreamt one night that a person offered her a red and a white rose, and that she chose the latter. On awaking she felt a burning pain in her arm, and by degrees, there arose there the figure of a rose perfect in form and colour. It was rather raised above the skin. The mark increased in intensity till the eighth day, after which it faded away, and by the fourteenth was no longer perceptible.

A letter from Moscow, addressed to Dr. Kerner, in consequence of reading the account of the Nun of Dulmen, relates a still more extraordinary case. At the time of the French

invasion, a Cossack having pursued a French-
man into a *cul de sac,* an alley without an out-
let, there ensued a terrible conflict between
them, in which the latter was severely wounded.
A person who had taken refuge in this close
and could not get away, was so dreadfully
frightened, that when he reached home, there
broke out on his body the very same wounds
that the Cossack had inflicted on his enemy.

The signatures of the fœtus are analogous
facts ; and if the mind of the mother can thus
act on another organism, why not the minds of
the saints, or of Catherine Emmerich, on their
own. From the influence of the mother on
the child, we have but one step to that asserted
to be possible, betwixt two organisms, not
visibly connected; for the difficulty therein
lies, that we do not see the link that con-
nects them, though, doubtless, it exists. Dr.
Blacklock, who lost his eyesight at an early
period, said, that, when awake, he distinguished persons by hearing and feeling them,
but when asleep, he had a distinct impression
of another sense. He then seemed to himself united to them by a kind of distant contact, which was effected by threads passing
from their bodies to his, which seems to be
but a metaphorical expression of the fact,

for, whether the connection be maintained
by an all-pervading ether, or be purely dy-
namic, that the interaction exists both
betwixt organic and inorganic bodies, is made
evident wherever there is sufficient excitability
to render the effects sensible. Till very
lately, the powers of the divining-rod were
considered a mere fable ; yet, that this power
exists, though not in the rod, but in the person that holds it, is now perfectly well
established. Count Tristan, who has written
a book on the subject, says, that about one
in forty have it, and that a complete course
of experiments has proved the phenomenon
to be electric. The rod seems to serve, in
some degree, the same purpose as the magical
mirror and conjurations, and it is, also, ser-
viceable in presenting a result visible to the
eye of the spectator. But, numerous cases
are met with, in which metals or water are
perceived beneath the surface of the earth,
without the intervention of the rod. A man,
called Bléton, from Dauphigny, possessed this
divining power in a remarkable degree, as did
a Swiss girl, called Katherine Beutler. She
was strong and healthy, and of a phlegmatic
temperament ; yet, so susceptible of these
influences, that, without the rod, she pointed

out and traced the course of water, veins of metal, coal beds, salt mines, &c. The sensations produced were sometimes on the soles of her feet, sometimes on her tongue, or in her stomach. She never lost the power wholly, but it varied considerably in intensity at different times, as it did with Bleton. She was also rendered sensible of the bodily pains of others, by laying her hand on the affected part, or near it, and she performed several magnetic cures.

A person now alive, named Dussange, in the Maçonnés, possesses this power. He is a simple honest man, who can give no account of his own faculty. The Abbés Chatelard and Paramelle can also discover subterraneous springs; but they say that it is effected by means of their geological science. Monsieur D., of Cluny, however, found the faculty of Dussange much more to be relied on. The Greeks and Romans made hydroscopy an art; and there are works alluded to as having existed on this subject; especially one by Marcellus. The caduceus of Mercury, the wand of Circe, and the wands of the Egyptian sorcerers, show that the wand or rod was always looked upon as a symbol of divination. One of the most remarkable instances

of the use of the divining-rod, is that of
Jacques Aymar.

On the 5th July, 1692, a man and his wife,
were murdered in a cellar at Lyons, and
their house was robbed. Having no clue
whatever to the criminal, this peasant, who
had the reputation of being able to discover
murderers, thieves, and stolen articles by
means of the divining rod, was sent for from
Dauphigny. Aymar undertook to follow the
footsteps of the assassins, but he said he
must first be taken into the cellar where
the murder was committed. The Procurator
Royal conducted him thither, and they gave
him a rod out of the first wood that came
to hand. He walked about the cellar but
the rod did not move till he came to the
spot where the man had been killed. Then
Aymar became agitated, and his pulse beat
as if he were in a high fever, and all these
symptoms were augmented when he approached the spot on which they had found
the body of the woman. From this, he, of
his own accord, went into a sort of shop
where the robbery had been committed; from
thence he proceeded into the street, tracing the
assassin, step by step, first to the court of the
Archbishop's palace, then out of the city and

along the right side of the river. He was
escorted all the way by three persons appointed for the purpose, who all testified that
sometimes he detected the traces of three
accomplices, sometimes only of two. He led
the way to the house of a gardener, where he
insisted that they had touched a table and one
of three bottles that were yet standing upon it.
It was at first denied; but two children of
nine or ten years old, said, that three men had
been there, and had been served with wine in
that bottle. Aymer then traced them to
the river where they had embarked in a
boat; and what is very extraordinary, he
tracked them as surely on the water as on the
land. He followed them wherever they had
gone ashore, went straight to the places they
had lodged at, pointed out their beds, and
the very utensils of every description that
they had used. On arriving at Sablon, where
some troops were encamped, the rod and his
own sensations satisfied him that the assassins
were there; but fearing the soldiers would ill-
treat him, he refused to pursue the enterprise
further, and returned to Lyons. He was, however, promised protection, and sent back by
water, with letters of recommendation. On
reaching Sablon, he said they were no longer

there, but he tracked them into Languedoc, entering every house they had stopped at, till he at length reached the gate of the prison, in the town of Beaucaire, where he said one of them would be found. They brought all the prisoners before him, amounting to fifteen; and the only one his rod turned on, was a little Bossu, or deformed man, who had just been brought in for a petty theft. He then ascertained that the two others had taken the road to Nimes, and offered to follow them; but as the man denied all knowledge of the murder, and declared he had never been at Lyons, it was thought best that they should return there; and as they went the way they had come, and stopped at the same houses, where he was recognized, he at length confessed that he had travelled with two men who had engaged him to assist in the crime. What is very remarkable, it was found necessary that Jacques Aymar should walk in front of the criminal; for, when he followed him, he became violently sick. From Lyons to Beaucaire is forty-five miles.

As the confession of the *Bossu* confirmed all Aymar had asserted, the affair now created an immense sensation, and a great variety of experiments were instituted, every

one of which proved perfectly satisfactory.
Moreover, two gentlemen, one of them the
Controller of the Customs, were discovered
to possess this faculty, though in a minor
degree. They now took Aymar back to
Beaucaire, that he might trace the other two
criminals, and he went straight again to the
prison gate, where he said, that now another
would be found. On enquiry however, it
was discovered that a man had been there to
enquire for the *Bossu*, but was gone again,
He then followed them to Toulon, and finally
to the frontier of Spain, which set a limit to
further researches. He was often so faint and
overcome with the effluvia, or whatever it was
that guided him, that the perspiration streamed
from his brow, and they were obliged to
sprinkle him with water to prevent his fainting.

He detected many robberies in the same
way. His rod moved whenever he passed
over metals or water, or stolen goods ; but he
found that he could distinguish the track of a
murderer from all the rest, by the horror and
pain he felt. He made this discovery accidentally as he was searching for water. They
dug up the ground, and found the body of a
woman that had been strangled.

I have myself met with three or four persons

in whose hands the rod turned visibly; and there are numerous very remarkable cases recorded in different works. In the Hartz, there is a race of people who support themselves entirely by this sort of divination; and as they are paid very highly, and do nothing else, they are generally extremely worthless and dissipated.

The extraordinary susceptibility to atmospheric changes in certain organisms, and the faculty by which a dog tracks the foot of his master, are analogous facts to those of the divining-rod. Mr. Boyle mentions a lady who always perceived if a person that visited her came from a place where snow had lately fallen. I have seen one, who if a quantity of gloves are given her can tell to a certainty to whom each belongs; and a particular friend of my own on entering a room, can distinguish perfectly who has been sitting in it; provided these be persons he is familiarly acquainted with. Numerous extraordinary stories are extant respecting this kind of faculty in dogs.

Doubtless, not only our bodies, but all matter, sheds its atmosphere around it; the sterility of the ground where metals are found, is notorious; and it is asserted that, to some persons, the vapours that emanate from below, are visible: and that, as the height of the mountains round a lake furnishes a measure of its depth, so does the height to which these vapours ascend, show how far below the surface the mineral treasures or the water lie. The effect of metals on somnambulic persons is well known to all who have paid any attention to these subjects; and surely may be admitted, when it is remembered that Humbold has discovered the same sensibility in Zoophytes, where no traces of nerves could be detected; and many years ago Frascatorius asserted that symptoms resembling apoplexy were sometimes induced by the proximity of a large quantity of metal. A gentleman is mentioned who could not enter the Mint at Paris, without fainting. In short so many well attested cases of idiosyncratic sensibilities exist, that we have no right to reject others because they appear incomprehensible.

Now, we may not only easily conceive, but we know it to be a fact, that fear, grief, and other detrimental passions, vitiate the secretions,* and augment transpiration; and it is quite natural to suppose, that where a crime

* In the "Medical Annals," a case is recorded of a young lady whose axillary excretions were rendered so offensive by the fright and horror she had experienced in seeing some of her relations assassinated in India, that she was unable to go into society.

has been committed which necessarily aroused a number of turbulent emotions, exhalations perceptible to a very acute sense, may for some time hover over the spot; whilst the anxiety, the terror, the haste, in short, the general commotion of system, that must accompany a murderer in his flight, is quite sufficient to account for his path being recognizable by such an abnormal faculty; "For the wicked flee, when no man pursueth." We also know that a person perspiring with open pores, is more susceptible than another to contagion; and we have only to suppose the pores of Jacques Aymar so constituted as easily to imbibe the emanations shed by the fugitive, and we see why he should be affected by the disagreeable sensations he describes. The disturbing effect of odours on some persons, which are quite innoxious to others, must have been observed by everybody. Some people do actually almost " die of a rose in aromatic pain." Boyle says that, in his time, many physicians avoided giving drugs to children, having found that external applications, to be imbibed by the skin, or by respiration, were sufficient ; and the Homeopaths occasionally use the same means now. Sir Charles Bell told me, that Mr. F., a gentle-

man well known in public life, had only to
hold an old book to his nose, to produce all
the effects of a cathartic. Elizabeth Okey was
oppressed with most painful sensations when
near a person whose frame was sinking.
Whenever this effect was of a certain intensity, Dr. Elliotson observed that the patient
invariably died.

Herein lies the secret of Amulets and Talis-
mans, which grew to be a vain superstition,
but in which, as in all popular beliefs, there
was a germ of truth. Somnambulic persons
frequently prescribe them; and absurd as it
may seem to many, there are instances in
which their efficacy has been perfectly established, be the interpretation of the mystery
what it may. In a great plague, which occurred
in Moravia, a physician, who was constantly
amongst the sufferers, attributed the complete
immunityof himself and his family to their
wearing amulets, composed of the powder of
toads ; " which," says Boyle, " caused an emʼa-
nation adverse to the contagion." A Dutch
physician mentions that in the great plague at
Nimeguen, the pest seldom attacked any house
till they had used soap in washing their linen.
Wherever this was done, it appeared immediately.

In short, we are the subjects, and so is everything around us, of all manner of subtle and inexplicable influences : and if our ancestors attached too much importance to these ill-understood arcana of the night side of nature, we have attached too little. The sympathetic effects of multitudes on each other, of the young sleeping with the old, of magnetism on plants and animals, are now acknowledged facts: may not many other asserted phenomena that we yet laugh at, be facts also, though probably too capricious in their nature—by which I mean, depending on laws beyond our apprehension—to be very available? For I take it, that as there is no such thing as chance, but all would be certainty if we knew the whole of the conditions, so no phenomena are really capricious and uncertain : they only appear so to our ignorance and shortsightedness.

The strong belief that formerly prevailed in the efficacy of sympathetic cures, can scarcely have existed, I think, without some foundation : nor are they a whit more extraordinary than the sympathetic falling of pictures and stopping of clocks and watches, of which such numerous well-attested cases are extant, that several learned German physiologists of the present day pronounce the thing indisputable. I have

myself heard of some very perplexing in
stances.

Gaffarillus alludes to a certain sort of magnet, not resembling iron, but of a black and
white colour, with which, if a needle or knife
were rubbed, the body might be punctured or
cut without pain. How can we know that this
is not true? Jugglers who slashed and cau-
terised their bodies for the amusement of the
public were supposed to avail themselves of
such secrets.

How is it possible for us, either, to imagine
that the numerous recorded cases of the *Blood
Ordeal*, which consisted in the suspected
assassin touching the body of his victim, can
have been either pure fictions or coincidences?
Not very long ago, an experiment of a
frightful nature is said to have been tried in
France on a somnambulic person, by placing
on the epigastric region a vial filled with the
arterial blood of a criminal just guillotined.
The effect asserted to have been produced, was
the establishment of a rapport between the
somnambule and the deceased, which endan-
gered the life of the former.

Franz von Baader suggests the hypothesis of
a *vis sanguinis ultra mortem*, and supposes that
a rapport or *communio vitæ* may be established

betwixt the murderer and his victim; and he conceives the idea of this mutual relation to be the true interpretation of the sacrificial rites common to all countries, as also of the *Blutschuld*, or the requiring blood for blood.

With regard to the blood ordeal, the following are the two latest instances of it recorded to have taken place in this country, they are extracted from "Hargrave's State Trials" :—

"Evidence having been given with respect to the death of Jane Norkott, an ancient and grave person, minister of the parish in Hertfordshire, where the murder took place, being sworn, deposed that, the body being taken up out of the grave, and the four defendants being present, were required, each of them, to touch the dead body. Okeman's wife fell upon her knees, and prayed God to show token of her innocency. The appellant did touch the body, whereon the brow of the deceased, which was before of a livid and carrion colour, began to have a dew, or gentle sweat on it, which increased by degrees till the sweat ran down in drops on the face, the brow turned to a lively and fresh colour, and the deceased opened one of her eyes and shut it again, and this opening the eye was done three several times;

she likewise thrust out the ring, or marriage finger, three times, and pulled it in again, and blood dropped from the finger on the grass.

"Sir Nicholas Hyde, the chief justice, seeming to doubt this evidence, he asked the witness, who saw these things besides him, to which he, the witness, answered, 'My Lord, I cannot swear what others saw, but I do believe the whole company saw it; and if it had been thought a doubt, proof would have been made, and many would have attested with me. My lord,' added the witness, observing the surprise his evidence awakened, ' I am minister of the parish, and have long known all the parties, but never had displeasure against any of them, nor they with me, but as I was minister. The thing was wonderful to me, but I have not interest in the matter, except as called on to testify to the truth. My Lord, my brother, who is minister of the next parish, is here present, and I am sure saw all that I have affirmed.' "

Hereupon, the brother, being sworn, he confirmed the above evidence in every particular, and the first witness added that, having dipped his finger into what appeared to be blood, he felt satisfied that it was really so. It is to be observed, that this extraordinary circumstance must have occurred, if it occurred at all, when the body had been upwards of a month dead; for it was taken up in consequence of various rumours implicating the prisoners, after the coroner's inquest had given in a verdict of *felo de se*. On their first trial, they were acquitted, but, an appeal being brought, they were found guilty and executed. It was on this latter occasion that the above strange evidence was given, which, being taken down at the time by Sir John Maynard, then Serjeant-at-Law, stands recorded, as I have observed, in Hargrave's edition of " State Trials."

The above circumstances occurred in the year 1628, and in 1688 the blood ordeal was again had recourse to in the trial of Sir Philip Stansfield for parricide, on which occasion the body had also been buried, but for a short time. Certain suspicions arising, it was disinterred and examined by the surgeons, and from a variety of indications, no doubt remained that the old man had been murdered, nor that his son was guilty of his death. When the body had been washed and arrayed in clean linen, the nearest relations and friends were desired to lift it and replace it in the coffin ; and when Sir Philip placed his hand under it, he suddenly drew it back, stained with blood, exclaiming, "Oh, God!" and letting the body fall, he cried, "Lord have mercy upon me!" and went and bowed himself over a seat in the church, in which the corpse had been inspected. Repeated testimonies are given to this circumstance in the course of the trial; and it is very remarkable that Sir John Dalrymple, a man of strong intellect, and wholly free from superstition, admits it as an established fact in his charge to the jury.

In short, we are all, though in different degrees, the subjects of a variety of subtle influences, which, more or less, neutralise each other, and many of which therefore we never observe; and frequently when we do observe the effects, we have neither time nor capacity for tracing the cause; and when in more susceptible organisms such effects are manifested, we content ourselves with referring the phenomena to disease or imposture. The exemption, or the power, whichever it may be, by which certain persons or races are enabled to handle venemous animals with impunity, is a subject that deserves much more attention than it has met with; but nobody thinks of investigating secrets that seem rather curious than profitable; besides which, to believe these things, implies a reflection on one's sagacity. Yet every now and then, I hear of facts so extraordinary, which come to me from undoubted authority, that I can see no reason in the world for rejecting others that are not much more so. For example, only the other day, Mr. B. C., a gentleman well known in Scotland, who has lived a great deal abroad, informed me, that having frequently heard of the singular phenomenon to be observed by placing a scorpion and a mouse together under a glass, he at length tried the experiment; and the result perfectly established what he had been previously unable to believe. Both animals were evidently frightened; but the scorpion made the first attack, and stung the mouse, which defended itself bravely, and killed the scorpion. The victory, however, was not without its penalties; for the mouse swelled to an unnatural size, and seemed in danger of dying from the poison of its defeated antagonist, when it relieved itself and was cured, by eating the scorpion, which was thus proved to be an antidote to its own venom; furnishing a most interesting and remarkable instance of isopathy.

There is a religious sect in Africa, not far from Algiers, which eat the most venomous serpents alive, and certainly, it is said, without extracting their fangs. They declare they enjoy the privilege from their founder. The creatures writhe, and struggle between their teeth; but possibly if they do bite them, the bite is innocuous.

Then, not to mention the common expedients of extracting the poisonous fangs, or forcing the animal by repeated bitings to exhaust their venom, the fact seems too well established to be longer doubted, that there are persons in whom the faculty of charming, or in other words disarming serpents, is inherent, as the Psylli and Marsi of old; and the people mentioned by Bruce, Hasselquist, and Lempriere, who were themselves eye-witnesses of the facts they relate. With respect to the Marsi, it must be remembered, that Heliogabalus made their priests fling venomous serpents into the circus, when it was full of people, and that many perished by the bites of these animals, which the Marsi had handled with impunity. The modern charmers told Bruce, that their immunity was born with them; and it was established beyond a doubt, during the French expedition into Egypt, that these people go from house to house to destroy serpents, as men do rats in this country. They declare that some mysterious instinct guides them to the animals, whom they immediately seize with fury, and tear to pieces with their hands and teeth. The negroes of the Antilles can smell a serpent which they do not see, and of whose presence a European is quite insensible, and Madame Calderon de la Barca mentions in her letters from Mexico, some singular cases of exemption from the pernicious effects of venomous bites ; and further relates that in some parts of America, where rattle-snakes are extremely abundant, they have a custom of innoculating children with the poison, and that this is a preservative from future injury. This may or may not be true; but it is so much the fashion, in these days, to set down to the account of fable everything deviating from our daily experience, that travellers may repeat these stories for ages, before any competent person will take the trouble of verifying the report. However, taking the evidence alto-gether, it appears clear that there does exist, in some persons, a faculty of producing in these animals, a sort of numbness, or *engour-dissement*, which renders them for the time incapable of mischief; though of the nature of the power we are utterly ignorant, unless it be magnetic. The senses of animals, although generally resembling ours, are yet extremely different in various instances; and we know that many of them have one faculty or another, exalted to an intensity of which we have no precise conception. Galen asserted, on the authority of the Marsi and Psylli themselves, that they obtained their immunity by feeding on the flesh of venomous animals; but Pliny, Elian, Silius Italicus, and others, account for the privilege by attributing it to the use of some substance of a powerful nature, with which they rubbed their bodies, and most modern travellers incline to the same explanation; but if this were the elucidation of the mystery, I suspect it would be easily detected.

It is observable, that in all countries where a secret of this sort exists, there is always found some custom which may be looked upon as either the cause or the consequence of the discovery. In Hindostan, for example, in order to test the truth of an accusation, the cobra cappello is flung into a deep pot of earth, with a ring; and if the supposed criminal succeeds in extracting the ring without being bitten by the serpent, he is accounted innocent. So the sacred asps in Egypt inflicted death upon the wicked, but spared the good. Dr. Allnut mentions that he saw a negro in Africa touch the protruded tongue of a snake with the black matter from the end of his pipe, which he said was tobacco oil. The effects were as rapid as a shock of electricity. The animal never stirred again, but stiffened, and was as rigid and hard as if it had been dried in the sun.

It is related of Machamut, a Moorish king, that he fed on poisons till his bite became fatal and his saliva venomous. Cœlius Rhodiginus mentions the same thing of a woman who was thus mortal to all her lovers; and Avicenna mentions a man whose bite was fatal in the same way.

The boy that was found in the forest of Arden, in 1563, and who had been nourished by a she wolf, made a great deal of money for a short time after he was introduced to civilized life, by exempting the flocks and herds of the shepherds, from the peril they nightly run of being devoured by wolves. This he did by stroking them with his hands, or wetting them with his saliva, after which, they for some time enjoyed an immunity. His faculty was discovered from the circumstance of the beasts he kept never being attacked. It left him, however, when he was about fourteen, and the wolves ceased to distinguish him from other human beings.

However, my readers will, I think, ere now
have supped full with *wonders*, if not with
horrors, and it is time I should bring this book
to a conclusion. If I have done no more, I
trust I shall at least have afforded some amusement; but I shall be better pleased to learn
that I have induced any one, if it be *but* one,
to look upon life and death, and the mysteries
that attach to both, with a more curious and
enquiring eye than they have hitherto done.
I cannot but think that it would be a great step
if mankind could familiarise themselves with
the idea that they are spirits incorporated for a
time in the flesh; but that the dissolution of
the connection between soul and body, though
it changes the external conditions of the former, leaves its moral state unaltered. What a
man has made himself, he will be; his state is
the result of his past life, and his heaven or
hell are in himself. At death, we enter upon
a new course of life; and what that life shall
be, depends upon ourselves. If we have provided oil for our lamps, and fitted ourselves for
a noble destiny, and the fellowship of the great
and good spirits that have passed away, such
will be our portion; and if we have misused
our talent and sunk our souls in the sensual
pleasures or base passions of this world, we

shall carry our desires and passions with us, to make our torment in the other; or perhaps be tethered to the earth by some inextinguishable remorse or disappointed scheme, like those unhappy spirits I have been writing about; and that perhaps for hundreds of years; for although evidently freed from many of the laws of space and matter, whilst unable to leave the earth, they are still the children of time, and have not entered into eternity. It is surely absurd to expect that because our bodies have decayed and fallen away, or been destroyed by an accident, that a miracle is to be wrought in our favour, and that the miser's love of gold, or the profligate's love of vice, are to be immediately extinguished, and be superseded by inclinations and tastes better suited to his new condition! New circumstances do not so rapidly engender a new mind here, that we should hope they will do so there: more especially, as, in the first place, we do not know what facilities of improvement may remain to us; and in the second, since the law, that like seeks like, must be undeviating, the blind will seek the blind, and not those who could help them to light.

I think, too, that if people would learn to remember that they are spirits, and acquire

the habit of conceiving of themselves as individuals, apart from the body, that they would
not only be better able to realize this view of a
future life, but they would also find it much
less difficult to imagine, that since they belong
to the spiritual world, on the one hand, quite
as much as they belong to the material world,
on the other, that these extraordinary faculties
which they occasionally see manifested by
certain individuals, or in certain states, may
possibly be but faint rays of those properties
which are inherent in spirit, though temporarily obscured by its connection with the
flesh; and designed to be so, for the purposes
of this earthly existence. The most ancient
nations of the world knew this, although we
have lost sight of it, as we learn by the sacred
books of the Hebrews.

According to the Cabbalah " Mankind are
endowed by nature, not only with the faculty
of penetrating into the regions of the super-
sensuous and invisible, but also of working
magically above and below; or in the worlds
of light and darkness. As the Eternal fills the
world, sees, and is not seen, so does the soul—
N'schamach—fill the body, and sees without
being seen. The soul perceives that which the
bodily eye cannot. Sometimes, a man is seized

suddenly with a fear, for which he cannot
account, which is, because the soul descrys an
impending misfortune. The soul possesses also
the power of working with the elementary
matter of the earth, so as to annihilate one
form, and produce another. Even by the force
of imagination, human beings can injure other
things; yea, even to the slaying of a man. (The
new platonist, Paracelsus, says the same thing.)
The Cabbalah teaches, that there have in all
times, existed men endowed with powers, in a
greater or less degree, to work good or evil, for
to be a virtuoso in either, requires a peculiar
spiritual vigour; thence, such men as heroes
and priests in the kingdom of Tumah—(the
kingdom of the clean and unclean.) If a man
therefore sets his desires on what is godly, in
proportion as his efforts are not selfish, but
purely a seeking of holiness, he will be endowed
by the free grace of God, with supernatural
faculties; and it is the highest aim of existence,
that man should regain his connction with his
inward, original source, and exalt the material
and earthly into the spiritual. The highest
degree of this condition of light and spirit, is
commonly called the holy extacy," which is
apparently, the degree attained by the extatics of the Tyrol.

I am very far from meaning to imply, that it is our duty, or in any way desirable, that we should seek to bring ourselves into this state of holy extacy, which seems to involve some derangement of the normal relations betwixt the soul and body, but it is at least equally unwise in us to laugh at, or deny it, or its proximate conditions, where they really exist. It appears perfectly clear, that, as by giving ourselves up wholly to our external and sensuous life, we dim and obscure the spirit of God that is in us, so by annihilating, as far as in us lies, the necessities of the body, we may so far subdue the flesh as to loosen the bonds of the spirit, and enable it to manifest some of its inherent endowments. Ascetics and saints have frequently done this voluntarily, and disease, or a peculiar constitution, sometimes does it for us involuntarily; and it is far from desirable that we should seek to produce such a state by either means, but it *is* extremely desirable that we should avail ourselves of the instruction to be gained by the simple knowledge that such phenomena have existed and been observed in all ages; and that thereby our connection with the spiritual world may become a demonstrated fact to all who choose to open their eyes to it.

With regard to the cases of apparitions I have adduced, they are not, as I said before, one hundredth part of those I could have brought forward, had I resorted to a few of the numerous printed collections that exist in all languages.

Whether the view I acknowledge myself to take of the facts be or not the correct one, whether we are to look to the region of the psychical or the hyperphysical, for the explanation, the facts themselves are certainly well worthy of observation; the more so, as it will be seen that, although ghosts are often said to be out of fashion, such occurrences are, in reality, as rife as ever; whilst, if these shadowy forms be actually visitors from the dead, I think we cannot too soon lend an attentive ear to the tale their re-appearance tells us.

That we do not all see them, or that those who promise to come, do not all keep tryst, amounts to nothing. We do not know why they can come, nor why they cannot; and as for not seeing them, I repeat, we must not forget how many other things there are that we do not see; and since, in science, we know that there are delicate manifestations which can only be rendered perceptible to our organs, by the application of the most delicate electrometers, is it not reasonable to suppose, that there may exist certain susceptible or diseased organisms, which, judiciously handled, may serve as electrometers to the healthy ones?

As my book is designed as an enquiry, with a note of interrogation, I characteristically bid adieu to my readers.

C. C.

THE END.